Recent Trends in

Animal Behaviour

Recent Trends in
Animal Behaviour

Archana Ruhela
Malini Sinha

Oxford Book Company
Jaipur ▪ New Delhi

Published By

Oxford Book Company

Regd. Off. "Bony Residency", Gate No. 2
Opp. Tilak Public School, Vishweswariya Nagar,
Gopalpura Road, Jaipur - 302018, India
Ph: 0141 - 2761280, Telefax: 2761381
email - oxfordbookjpr@gmail.com
website - oxfordbookcompany.com

Delhi Office: 102, 1st Floor, Satyam House,
4327/3, Ansari Road, Darya Ganj,
New Delhi - 110002, India
Ph: 011-45652440

Distributed by
KEYA INTERNATIONAL PRESS LLP
E-mail: keyainternationalpressllp@gmail.com

Published in India, 2025

ISBN : 9789380179261

Typeset by Shivangi Computers, Jaipur

Printed and bound in India by Replika Press Pvt. Ltd.

Preface

Animal behaviour is the scientific study of the wild and wonderful ways in which animals interact with each other, with other living beings, and with the environment. It explores how animals relate to their physical environment as well as to other organisms, and includes topics such as how animals find and defend resources, avoid predators, choose mates and reproduce, and care for their young. Behaviours exist on a continuum from being largely pre-determined by genetics, to others which are mostly acquired through experience. Instinctive components are those that are exhibited without the benefit of, or need for, prior experience. They come to the fore when, for instance, newly hatched, inexperienced individuals of a particular species of spider construct their webs with little variation, despite the lack of a tutor. Learned behaviours are those that are acquired through interactions with the surrounding world, the "nurture" part of the nature vs. nurture equation.

The present book explores the mechanisms and evolution of animal behaviour, including neural, hormonal, and genetic substrates of behaviour; foraging; anti-predator defences; mating systems and sexual selection; social behaviour; communication; parental care; kin selection and recognition; and territoriality. This thoroughly up-to-date text shows how evolutionary biologists analyse all aspects of behaviour. It is distinguished by its balanced treatment of both the underlying mechanisms and evolutionary causes of behaviour, and stresses the utility of evolutionary theory in unifying the different behavioural disciplines. Important concepts are explained by reference to key illustrative studies, which are described in sufficient detail to help students appreciate the role of the scientific process in producing research discoveries. The writing style is clear and engaging: beginning students have no difficulty following the material, despite the strong conceptual orientation of the text. This is essential reading for degree-level students following modular programs in biology, zoology, marine biology, and psychology.

Archana Ruhela
Malini Sinha

Contents

1

Introduction to Animal Behaviour

Animal behaviour is the scientific study of the wild and wonderful ways in which animals interact with each other, with other living beings, and with the environment. It explores how animals relate to their physical environment as well as to other organisms, and includes topics such as how animals find and defend resources, avoid predators, choose mates and reproduce, and care for their young. Behaviours exist on a continuum from being largely pre-determined by genetics to others which are mostly acquired through experience.

Instinctive components are those that are exhibited without the benefit of, or need for, prior experience. They come to the fore when, for instance, newly hatched, inexperienced individuals of a particular species of spider construct their webs with little variation, despite the lack of a tutor. Learned behaviours are those that are acquired through interactions with the surrounding world, the "nurture" part of the nature vs. nurture equation.Web making by spiders is an example of a genetically determined or instinctive behaviour. There is little variation between individuals in how they construct the web and it is constructed similarly each time they do it. Ethologists (people who study behaviour) call such a behaviour a fixed action pattern. Fixed action patterns do not require learning or prior experience for their expression. They can, however, be very complex. For example it has been noted that a cocoon-spinning spider performs over 6000 individual movements in a virtually identical fashion each time it prepares and closes its cocoon.

Fixed action patterns cannot be identified solely because they are highly stereotypic and species-specific. For example, songs of bird species fit those criteria but are actually learned behaviours, not instinctive behaviours. Birds deprived of the experience of hearing the song do not produce the characteristic song pattern. Deprivation experiments in which animals are raised without parents, or contact with their own species, have typically been used to help distinguish a behaviour that is a fixed action pattern from those which are learned.

The lack of expression of a behaviour in a deprivation study does not necessarily mean that it is not genetically determined. It may mean that the requisite stimuli are absent. Stimuli to elicit the expression of fixed action patterns are called releasers. Also a fixed action pattern may not be displayed because the animal is not in the appropriate physiological or developmental state. For example courtship behaviours are not shown by pre-pubertal animals even if the appropriate releasers are present.

Cognition is the ability to think. When applied in the context of animal behaviour it refers to the ability of an animal to be aware of and make judgements about its environment. Are animals conscious of themselves and their place in the world? Do they feel pain, pleasure and sadness in the same way as humans? Are the differences between animals and humans in cognitive ability absolute, as believed by Descartes, or a matter of degree? Professor Donald Griffin of Princeton has been a proponent of the concept of cognitive ethology. This views conscious thinking as an inherent and essential part of the behaviour of many non-human animals. Conscious thinking has been demonstrated in non-human primates but it is a very difficult subject to study with scientific rigor. As we have discussed above very many complex behaviours are actually genetically determined or learned by simple associations that do not need cognition. Difficult though they are to study the answers to questions about animal cognition may profoundly affect our views on the treatment of animals and their welfare.

Communication among animals relies upon their abilities to perceive sensory information. This may be visual, auditory or olfactory. Domestic animals perceive the world in a different fashion to us so may respond differently also. An obvious example of seeing the world differently relates to eye placement. Horses are set laterally, providing a wide range of monocular vision (215 degrees) but a small 60-70 degree arc of binocular vision. In contrast, the cat has a much smaller arc of vision (around 180 degree) because of the placement of the eyes to the front but more than half of this is binocular vision. When the relative visual acuity of domestic animals is compared, the ranking is pig, sheep, cattle, dogs and horses. All the domestic animals have some ability to discriminate colours and most have superior night vision to humans.

In the horse, for example, the ear position can tell a lot about the animals disposition. Ears pointed back generally denote aggression and the flatter the ears to the head the greater the aggression. Another visual expression in male horses and ruminants is the Flehmen response. This is a curl of the lip when the urine of a female in estrus is smelled. In the cat, a high tail is a greeting or sign of being curious. Raised hackles, and lips drawn back in the dog denotes aggression; if the ears are flat against the head and the tail is between the legs this signifies fear. Pigs and horses show grooming behaviour, as do many of the monkeys and primates. Subordinate pigs groom dominant ones. Horses tend to groom horses of comparable rank. Cattle and cats spend long periods of time self-grooming but little time grooming others. Subordinate cattle will lick dominant cattle.

Again there are differing ranges of hearing, as well as different acuities. Sheep and dogs can discriminate higher frequencies than humans. Cats range of hearing is similar to humans. Pigs have perhaps the most complex set of domestic animal vocal sounds—more than 20 distinct types have been identified. Horses, cats and dogs also have many sounds but those of cattle and sheep have been little studied. The categories of calls include greeting, distress, separation, excitement, aggression, fear, pain and so on.

Olfactory senses is perhaps the most important sense of domestic animals in terms of communication. Dogs probably have the greatest olfactory ability of the domestic species. They are able to detect many compounds at 1/100 the concentration of humans and for many weeks after they were placed. Odors and pheremones are very important triggers for domestic animal behaviour. Olfactory clues and scent serve to mark territory, show the way home, and distinguish individuals. Urine and feces are powerful means of olfactory communication. Cats and dogs also have anal sacs which are additional scent organs. Olfactory stimuli play a major part in heat detection in cattle, and for distinguishing animals.

SCOPE OF ANIMAL BEHAVIOUR

Niko Tinbergen's legacy as a co-founder of Ethology and exceptionally gifted experimentalist include the phrasing of four different types of questions that can be asked of any animal behaviour.

Proximate Explanations

Neural, Endocrine, and Genetic Mechanisms

Causes that explain how a behaviour is elicited and coordinated. What are the anatomical or physiological mechanisms that underly the behaviour? The behaviour is present because the nervous system makes it happen. It includes a search for the impact of genetic components, physiological mechanisms, environmental conditions needed, and hormones on behaviour.

Ontogeny, Development

How does the behaviour develop and change during the lifespan of a single individual? The behaviour is present because the individual developed an ability to do it this way.

Ultimate Explanations

Survival Value

Explanations are phrased as ultimate causes that explain why a behaviour has evolved.

Selection ought to favor individuals with behaviours that help it increase its reproductive success. The behaviour is present because the individual inherited them from parents who were historically successful with this behaviour. This includes a search for functional significance or adaptiveness of behaviour.

Phylogeny, Evolution

Phylogeny is the study of evolutionary relatedness among groups of organisms as It treats a species as a group of lineage-connected individuals over time. It attempts to historically characterise the branching processes in which breeding populations have changed over time. The ethologist's view holds that behavioural characteristics, which receive at least a partial genetic contribution, and can be viewed in a phylogenetic perspective like any other morphological trait. The behaviour is present because the individual is an offspring from ancestors that had the behaviour.

HISTORY OF ANIMAL BEHAVIOUR

Our early ancestors critically depended on a detailed understanding of animal behaviour. Specifically, knowledge of habitat preference, movement patterns, and sensory biology of prey as well as predators were essential. Rock paintings, figurines, and carvings, depicting scenes from daily life and religious rituals, commonly feature detailed likenesses of many kinds of animals.

Holes in Neolithic skulls illustrate that patients dating back to such early times have been subjected to head surgery, presumably with the aim of curing headaches or mental disorders. Sumerian descriptions show the use of Opium, derived from poppy plants, to produce euphoriant effects.

A common theme contends that the world was populated by many peoples who were subsequently transformed into different kinds of animals. Thus all animals and humans share the presence of a soul. The close bond between them is shown in frequent rituals where the behaviour of different animals is simulated. Imagery of animals is common. The Great Serpent Mound represents an effigy mound located on a plateau along Ohio Brush Creek in Adams County, Ohio. Conforming to the topography, the snake's head approaches a cliff point, the snake then winds back and forth for more than 200 meters and ends with a triple coiled tail.

Ancient Egyptian societies showed a keen interest in the temperaments and sensory abilities of different animals and ascribed them to individual deities.

Ancient Egyptians had a highly developed view of the afterlife with elaborate rituals for preparing the body and soul for a peaceful existence after death. Beliefs about the soul and its needs led to numerous ceremonies for the preservation of the body. The brain apparently was not considered an essential element for eternity, as it was treated with a distinct lack of dignity. It was often spooned into small pieces and removed through the nose, then unceremoniously discarded.

Ancient Greek Period

Plato held the notion that the human mind as an entity pre-existed somewhere in the heavens, before being sent down to join a body on earth. Hippocrates believed that the brain was not only involved with sensation but was also the seat of intelligence. Aristotle viewed human nature as constituting formed habits and a desire for happiness. His writings contain the first references to the concept of a tabula rasa. This refers to the thesis that individuals are born with no innate or built-in mental content, in a word, "blank", and that their entire resource of knowledge is built up gradually from their experiences and sensory perceptions of the outside world. Herophilus studies the nervous system and distinguishes between sensory nerves and motor nerves.

Roman Empire

As a physician in a gladiator school, Galen gained a unique view of the behavioural deficits that resulted from acute head trauma. Not surprisingly he wrote extensively on the connection between paralysis and severance of the spinal cord. His many audacious operations, including brain and eye surgery, were not again used for almost two millennia.

Middle Ages

Naturalists and philosophers focused on classical texts and, in particular, on the Bible in Latin. Accepted were Aristotle's views on all matters of science, including psychology. As knowledge of Greek declined, the West found itself cut off from its Greek philosophical and scientific roots. General levels of literacy further decreased as learning became largely restricted to monastic and cathedral schools that were primarily aimed at training clergy. The Church exerted an overpowering role in philosophical thought and nontheological scholarship was rare. Despite this many influences on the study of behaviour in later periods began to emerge.

Renaissance Period

The principal features of the Renaissance (14th-16th century) were the revival of learning based on classical sources and the broad advancements of science. It pioneered scientific study through deductive reasoning. René Descartes supplied a foundation for the pursuit of knowledge.

— Unity of all sciences through deductive reasoning

— Never accept anything as true that is not known clearly to be such
— Divide problems into as many parts as possible, and proceed from the simplest to the hardest

He argues that he perceives his body through the use of the senses, which often proven unreliable. So he concludes that thinking is his essence as it is the only thing about him that cannot be doubted. Descartes defines "thought" as every activity of which a person is immediately conscious. In his Mind/body dualism he viewed the body as a machine that follows the laws of physics. In contrast, he regarded the mind (or soul) as a nonmaterial entity that is not subject to such physical constraints.

With the advent of the scientific revolution, medieval worldviews steadily gave way to increasingly analytical methodologies and a weakening of vitalist traditions. Although medical thought continued to be guided by the occult, critical thinking began to chip away at the long-held notions based on Galen. Hippocrates' views are revived and Renaissance artists such as Leonardo da Vinci conducted detailed studies of human anatomy in order to better portray the human form. As Andreas Vesalius corrects Greek medical errors, he revolutionises medicine. William Harvey applies scientific methodology to human physiology and solves the human circulatory system. Paracelsus rejects occultism and pioneers the use of chemicals and minerals in medicine.

Enlightenment Period

Thinkers of the Age of Reason perceive their efforts as bringing to an end previous periods of irrationality, superstition, and tyranny. John Locke's philosophy revives Aristiotles theory of a tabula rasa, where the (human) mind is at birth a "blank slate" without rules for processing data, and where data is added and rules for processing are formed solely by one's sensory experiences. The concept of a blank mind at birth emphasised the individual's freedom to author one's own human character. Moreover, as members of ones society will become what society make of them, it becomes essential to assure universal access to learning opportunities, medical care, and possibilities for personal development.

STUDIES LEAD TO THE DEVELOPMENT ON ANIMAL BEHAVIOUR

Psychophysics

Psychophysicists, such as Gustav Theodor Fechner and Wilhelm Wundt, aimed to relate matter to mind via a study of inner experiences or feelings towards defined physical stimuli. Psychophysicists explicitly rejected the concepts of vitalists, i.e. that the functions of a living organism are due to a vital principle distinct from and not expliciable by physicochemical forces. The main focus lay on the way subjects perceive experimental stimuli that can be objectively measured, such as pure tones varying in intensity, or lights varying in luminance. A broad range of senses has been studied, including vision, hearing, touch, taste, smell, or the sense of time.

Psychophysics can be used to produce scales of human experience of various aspects of physical stimuli. Take for an example the physical stimulus of frequency of sound. Frequency of a sound is measured in hertz, cycles per second. But human experience of the frequencies of sound is not the same as the frequencies. For one thing, there is a frequency below which no sounds can be heard, no matter how intense they are (around 20 Hz depending on the individual) and there is a frequency above which no sounds can be heard, no matter how intense they are (around 20,000 Hz, again depending on the individual). For another, doubling the frequency of a sound (e.g., from 100 Hz to 200 Hz) does not lead to a doubling of experience. The perceptual experience of the frequency of sound is called pitch, and it is measured by psychophysicists in mels. This information can be useful in, for example, producing soundtracks for movies that most efficiently utilise the available range of frequencies.

More analytical approaches allow the use of psychophysical methods to study neurophysiological properties and sensory processing mechanisms. This is of particular importance in human research, where other (more invasive) methods are not used due to ethical reasons.

Psychophysicists usually employ experimental stimuli that can be objectively measured, such as pure tones varying in intensity, or lights varying in luminance. All the senses have been studied: vision, hearing, touch (including skin and enteric perception), taste, smell, and the sense of time. Regardless of the sensory domain, there are three main areas of investigation: absolute thresholds, discrimination thresholds, and scaling.

A threshold (or limen), is the point of intensity at which the participant can just detect the presence of, or difference in, a stimulus. Stimuli with intensities below the threshold are considered not detectable (hence: sub-liminal), however stimuli at values close to threshold will often be detectable some proportion of the time. Due to this, a threshold is considered to be the point at which a stimulus, or change in a stimulus, is detected some proportion p of the time. There are two kinds of thresholds: absolute and difference.

An absolute threshold is the level of intensity of a stimulus at which the subject is able to detect the presence of the stimulus some proportion of the time (a p level of 50% is often used). An example of an absolute threshold is the number of hairs on the back of one's hand that must be touched before it can be felt - a participant may be unable to feel a single hair being touched, but may be able to feel two or three as this exceeds the threshold. Absolute threshold is also often referred to as detection threshold.

A difference threshold is the magnitude of the difference between two stimuli of differing intensities that the participant is able to detect some proportion of the time (again, 50% is often used). To test this threshold, several different methods are used. The subject may be asked to adjust one stimulus until it is perceived as the same as the other, may be asked to describe the magnitude of the difference between two stimuli, or may be asked to detect a stimulus against a background.

In discrimination experiments, the experimenter seeks to determine at what point the difference between two stimuli, such as two weights or two sounds, is detectable. The subject is presented with one stimulus, for example a weight, and is asked to say whether another weight is heavier or lighter (in some experiments, the subject may also say the two weights are the same). At the point of subjective equality (PSE), the subject perceives the two weights to be the same. The just noticeable difference (JND), or difference limen (DL), is the difference in stimuli that the subject notices some proportion p of the time (50% is usually used for p).

Absolute and difference thresholds are sometimes considered similar because there is always background noise interfering with our ability to detect stimuli, however study of difference thresholds still occurs, for example in pitch discrimination tasks

In psychophysics, experiments seek to determine whether the subject can detect a stimulus, identify it, differentiate between it and another stimulus, and describe the magnitude or nature of this difference. Psychophysical experiments have traditionally used three methods for testing subjects' perception in stimulus detection and difference detection experiments: the method of limits, the method of constant stimuli, and the method of adjustment.

In ascending method of limits, some property of the stimulus starts out at a level so low that the stimulus could not be detected, then this level is gradually increased until the participant reports that they are aware of it. For example, if the experiment is testing the minimum amplitude of sound that can be detected, the sound begins too quietly to be perceived, and is made gradually louder. In the descending method of limits, this is reversed. In each case, the threshold is considered to be the level of the stimulus property at which the stimuli is just detected.

In experiments, the ascending and descending methods are used alternately and the thresholds are averaged. A possible disadvantage of these methods is that the subject may become accustomed to reporting that they perceive a stimulus and may continue reporting the same way even beyond the threshold (the error of habituation). Conversely, the subject may also anticipate that the stimulus is about to become detectable or undetectable and may make a premature judgment (the error of expectation). To avoid these potential pitfalls, Georg von Bekesy introduced the staircase procedure in 1960 in his study of auditory perception. In this method, the sound starts out audible and gets quieter after each of the subject's responses, until the subject does not report hearing it. At that point, the sound is made louder at each step, until the subject reports hearing it, at which point it is made quieter in steps again. This way the experimenter is able to "zero in" on the threshold. Instead of being presented in ascending or descending order, in the method of constant stimuli the levels of a certain property of the stimulus are not related from one trial to the next, but presented randomly. This prevents the subject from being able to predict the level of the next stimulus, and therefore reduces errors of habituation and expectation. The subject again reports whether he or she is able to detect the stimulus.

The method of adjustment asks the subject to control the level of the stimulus, instructs them to alter it until it is just barely detectable against the background noise, or is the same as the level of another stimulus. This is repeated many times. This is also called the method of average error. Often, the classic methods of experimentation are argued to be inefficient. This is because, in advance of testing, the psychometric threshold is usually unknown and a lot of data has to be collected at points on the psychometric function that provide little information about its shape (the tails). Adaptive staircase procedures can be used such that the points sampled are clustered around the psychometric threshold. However, the cost of this efficiency, is that you do not get the same amount of information regarding the shape of the psychometric function as you can through classical methods. Despite this, it is still possible to estimate the threshold and slope by fitting psychometric functions to the obtained data, although estimates of psychometric slope are likely to be more variable than those from the method of constant stimuli (for a reasonable sampling of the psychometric function).

Staircases usually begin with a high intensity stimulus, that is easy to detect. The intensity is then reduced until the observer makes a mistake, at which point the staircase 'reverses' and intensity is increased until the observer responds correctly, triggering another reversal. The values for these 'reversals' are then averaged. There are many different types of staircase, utilising many different decision and termination rules. Step-size, up/down rules and the spread of the underlying psychometric function dictate where on the psychometric function they converge. Threshold values obtained from staircases can fluctuate wildly, so care must be taken in their design. Many different staircase algorithms have been modeled and some practical recommendations suggested by Garcia-Perez.

Psychoanalysis

Psychoanalysis is a family of psychological theories and methods based on the pioneering work of Sigmund Freud. As a result of talking with his patients, he came to believe that psychopathologies stemmed from culturally unacceptable, thus repressed and unconscious, desires and fantasies of a sexual nature. Moreover he stressed the presumed value of dreams as sources of insight into unconscious desires. The analyst's goal is thus aimed to discover connections among the unconscious components of patients' mental processes. This ought to help liberate the patient from unexamined or unconscious barriers of transference and resistance, that is, past patterns of relating that are no longer serviceable or that inhibit freedom.

Carl G. Jung's unique and broadly influential approach to psychology emphasised understanding the psyche through exploring the worlds of dreams, art, mythology, world religion and philosophy. Jung emphasised the importance of balance and harmony. He cautioned that modern humans rely too heavily on science and logic and would benefit from integrating spirituality and appreciation of the unconscious realm. His Collective unconscious refers to a human experience reservoir, a kind of knowledge, we are all born with but can never be directly conscious of. His Archetypes represent generic, idealised models of personalities with unlearned tendencies to experience things in a certain way.

Under the broad umbrella of psychoanalysis there are at least 22 different theoretical orientations regarding the underlying theory of understanding of human mentation and human development. The various approaches in treatment called "psychoanalytic" vary as much as the different theories do. In addition, the term refers to a method of studying child development.

Freudian psychoanalysis refers to a specific type of treatment in which the "analysand" (analytic patient) verbalizes thoughts, including free associations, fantasies, and dreams, from which the analyst formulates the unconscious conflicts causing the patient's symptoms and character problems, and interprets them for the patient to create insight for resolution of the problems.

The specifics of the analyst's interventions typically include confronting and clarifying the patient's pathological defenses, wishes and guilt. Through the analysis of conflicts, including those contributing to resistance and those involving transference onto the analyst of distorted reactions, psychoanalytic treatment can clarify how patients unconsciously are their own worst enemies: how unconscious, symbolic reactions that have been stimulated by experience are causing symptoms.

The various psychoses involve deficits in the autonomous ego functions (see above) of integration (organisation) of thought, in abstraction ability, in relationship to reality and in reality testing. In depressions with psychotic features, the self-preservation function may also be damaged (sometimes by overwhelming depressive affect). Because of the integrative deficits (often causing what general psychiatrists call "loose associations," "blocking," "flight of ideas," "verbigeration," and "thought withdrawal"), the development of self and object representations is also impaired. Clinically, therefore, psychotic individuals manifest limitations in warmth, empathy, trust, identity, closeness and/or stability in relationships (due to problems with self-object fusion anxiety) as well.

In patients whose autonomous ego functions are more intact, but who still show problems with object relations, the diagnosis often falls into the category known as "borderline." Borderline patients also show deficits, often in controlling impulses, affects, or fantasies – but their ability to test reality remains more or less intact. Adults who do not experience guilt and shame, and who indulge in criminal behaviour, are usually diagnosed as psychopaths, or, using DSM-IV-TR, antisocial personality disorder.

Panic, phobias, conversions, obsessions, compulsions and depressions (analysts call these "neurotic symptoms") are not usually caused by deficits in functions. Instead, they are caused by intrapsychic conflicts. The conflicts are generally among sexual and hostile-aggressive wishes, guilt and shame, and reality factors. The conflicts may be conscious or unconscious, but create anxiety, depressive affect, and anger. Finally, the various elements are managed by defensive operations – essentially shut-off brain mechanisms that make people unaware of that element of conflict. "Repression" is the term given to the mechanism that shuts thoughts out of consciousness. "Isolation of affect" is the term used for the mechanism that shuts sensations out of consciousness. Neurotic symptoms may occur with or without deficits in ego functions, object relations, and ego strengths. Therefore, it is not uncommon to encounter obsessive-compulsive schizophrenics, panic patients who also suffer with borderline personality disorder, etc.

Freudian theories argue that adult problems can be traced to unresolved conflicts from certain phases of childhood and adolescence. Freud, based on the data gathered from his patients early in his career, suspected that neurotic disturbances occurred when children were sexually abused in childhood (the so-called seduction theory). Later, Freud came to realise that, although child abuse occurs, that not all neurotic symptoms were associated with this. He realised that neurotic people often had unconscious conflicts that involved incestuous fantasies deriving from different stages of development. He found the stage from about three to six years of age (preschool years, today called the "first genital stage") to be filled with fantasies about marriage with both parents. Although arguments were generated in early 20th-century Vienna about whether adult seduction of children was the basis of neurotic illness, there is virtually no argument about this problem in the 21st century.

Many psychoanalysts who work with children have studied the actual effects of child abuse, which include ego and object relations deficits and severe neurotic conflicts. Much research has been done on these types of trauma in childhood, and the adult sequelae of those. On the other hand, many adults with symptom neuroses and character pathology have no history of childhood sexual or physical abuse. In studying the childhood factors that start neurotic symptom development, Freud found a constellation of factors that, for literary reasons, he termed the Oedipus complex (based on the play by Sophocles, Oedipus Rex, where the protagonist unwittingly kills his father Laius and marries his mother Jocasta). The shorthand term, "oedipal," (later explicated by Joseph Sandler in "On the Concept Superego" (1960) and modified by Charles Brenner in "The Mind in Conflict" (1982)) refers to the powerful attachments that children make to their parents in the preschool years. These attachments involve fantasies of marriage to either (or both) parent, and, therefore, competitive fantasies toward either (or both) parents. Humberto Nagera (1975) has been particularly helpful in clarifying many of the complexities of the child through these years.

The terms "positive" and "negative" oedipal conflicts have been attached to the heterosexual and homosexual aspects, respectively. Both seem to occur in development of most children. Eventually, the developing child's concessions to reality (that they will neither marry one parent nor eliminate the other) lead to identifications with parental values. These identifications generally create a new set of mental operations regarding values and guilt, subsumed under the term "superego." Besides superego development, children "resolve" their preschool oedipal conflicts through channeling wishes into something their parents approve of ("sublimations") and the development, during the school-age years ("latency") of age-appropriate obsessive-compulsive defensive maneuvers (rules, repetitive games).

Using the various analytic theories to assess mental problems, several particular constellations of problems are particularly suited for analytic techniques (see below) whereas other problems respond better to medicines and different interpersonal interventions. To be treated with psychoanalysis, whatever the presenting problem, the person requesting help must demonstrate a good capacity to organise thought (integrative function), good abstraction ability, and a reasonable ability to observe self and others. As well, they need to be able to have trust and empathy and they must be able to control emotion and urges. Potential patients must be in contact with reality, which excludes most psychotic patients with delusions, and they must feel some guilt and shame (this requirement excludes some criminals and sex offenders who do not feel remorse). Finally, a prospective patient must not be severely suicidal. If any of the above are faulty, then modifications of techniques, or completely different treatment approaches, must be instituted.

The more there are deficits of serious magnitude in any of the above mental operations (1-8), the more psychoanalysis as treatment is contraindicated, and the more medication and supportive approaches are indicated. In non-psychotic first-degree criminals, any treatment is often contraindicated. The problems treatable with analysis include: phobias, conversions, compulsions, obsessions, anxiety attacks, depressions, sexual dysfunctions, a wide variety of relationship problems (such as dating and marital strife), and a wide variety of character problems (for example, painful shyness, meanness, obnoxiousness, workaholism, hyperseductiveness, hyperemotionality, hyperfastidiousness). The fact that many of such patients also demonstrate deficits above makes diagnosis and treatment selection difficult.

The match between the analyst and the patient can be viewed as another contributing factor for the indication and contraindication for psychoanalytic treatment. The analyst decides whether the patient is suitable for psychoanalysis. This decision made by the analyst, besides made on the usual indications and pathology, is also based to a certain degree by the "fit" between analyst and patient. When analysts utilise concrete, semi-standardised procedures to evaluate patients' suitability for analytic treatment, their associations' "defined protocols," may include (semi-) structured interviews, personality tests, projective tests, and/or psychological questionnaires. An evaluation may include one or more other analysts' independent opinions and will include discussion of the patient's financial situation and insurances.

Behaviourism

This school of thought represents a movement in psychology that advocates the use of strict experimental procedures to study observable behaviours (or responses) in relation to environment (or stimuli). Behaviourists regard learning and environmental conditions as dominant over the effects of genetics and heredity. Concepts, such as learning or memory are explored using artificial tasks in strictly controlled environments.

Behaviourism comprises the position that all theories should have observational correlates but that there are no philosophical differences between publicly observable processes (such as actions) and privately observable processes (such as thinking and feeling). From early psychology in the 19th century, the behaviourist school of thought ran concurrently and shared commonalities with the psychoanalytic and Gestalt movements in psychology into the 20th century; but also differed from the mental philosophy of the Gestalt psychologists in critical ways. Its main influences were Ivan Pavlov, who investigated classical conditioning, Edward Lee Thorndike, John B. Watson who rejected introspective methods and sought to restrict psychology to experimental methods, and B.F. Skinner who conducted research on operant conditioning. In the second half of the twentieth century, behaviourism was largely eclipsed as a result of the cognitive revolution.

Skinner, who carried out experimental work mainly in comparative psychology from the 1930s to the 1950s, but remained behaviourism's best known theorist and exponent virtually until his death in 1990, developed a distinct kind of behaviourist philosophy, which came to be called radical behaviourism. He is credited with having founded a new version of psychological science, which has come to be called behaviour analysis or the experimental analysis of behaviour after variations on the subtitle to his 1938 work The Behaviour of Organisms: An Experimental Analysis Of Behaviour.

B.F. Skinner was influential in defining radical behaviourism, a philosophy codifying the basis of his school of research (named the Experimental Analysis of Behaviour, or EAB.) While EAB differs from other approaches to behavioural research on numerous methodological and theoretical points, radical behaviourism departs from methodological behaviourism most notably in accepting treatment of feelings, states of mind and introspection as existent and scientifically treatable. This is done by identifying them as something non-dualistic, and here Skinner takes a divide-and-conquer approach, with some instances being identified with bodily conditions or behaviour, and others getting a more extended 'analysis' in terms of behaviour. However, radical behaviourism stops short of identifying feelings as causes of behaviour. Among other points of difference were a rejection of the reflex as a model of all behaviour and a defense of a science of behaviour complementary to but independent of physiology. Radical behaviourism has considerable overlap with other western philosophical positions such as American pragmatism.

This essentially philosophical position gained strength from the success of Skinner's early experimental work with rats and pigeons, summarised in his books *The Behaviour of Organisms and Schedules of Reinforcement*. Of particular importance was his concept of the operant response, of which the canonical example was the rat's lever-press. In contrast with the idea of a physiological or reflex response, an operant is a class of structurally distinct but functionally equivalent responses. For example, while a rat might press a lever with its left paw or its right paw or its tail, all of these responses operate on the world in the same way and have a common consequence. Operants are often thought of as species of responses, where the individuals differ but the class coheres in its function—shared consequences with operants and reproductive success with species. This is a clear distinction between Skinner's theory and S-R theory.

Skinner's empirical work expanded on earlier research on trial-and-error learning by researchers such as Thorndike and Guthrie with both conceptual reformulations – Thorndike's notion of a stimulus-response 'association' or 'connection' was abandoned – and methodological ones – the use of the 'free operant', so called because the animal was now permitted to respond at its own rate rather than in a series of trials determined by the experimenter procedures. With this method, Skinner carried out substantial experimental work on the effects of different schedules and rates of reinforcement on the rates of operant responses made by rats and pigeons. He achieved remarkable success in training animals to perform unexpected responses, and to emit large numbers of responses, and to demonstrate many empirical regularities at the purely behavioural level. This lent some credibility to his conceptual analysis. It is largely his conceptual analysis that made his work much more rigorous than his peers, a point which can be seen clearly in his seminal work Are Theories of Learning Necessary? in which he criticises what he viewed to be theoretical weaknesses then common in the study of psychology. An important descendant of the experimental analysis of behaviour is the Society for Quantitative Analysis of Behaviour.

As Skinner turned from experimental work to concentrate on the philosophical underpinnings of a science of behaviour, his attention turned to human language with Verbal Behaviour and other language-related publications; Verbal Behaviour laid out a vocabulary and theory for functional analysis of verbal behaviour, and was strongly criticised in a review by Noam Chomsky. Skinner did not respond in detail but claimed that Chomsky failed to understand his ideas, and the disagreements between the two and the theories involved have been further discussed.

What was important for a behaviourist's analysis of human behaviour was not language acquisition so much as the interaction between language and overt behaviour. In an essay republished in his 1969 book Contingencies of Reinforcement, Skinner took the view that humans could construct linguistic stimuli that would then acquire control over their behaviour in the same way that external stimuli could. The possibility of such "instructional control" over behaviour meant that contingencies of reinforcement would not always produce the same effects on human behaviour as they reliably do in other animals. The focus of a radical behaviourist analysis of human behaviour therefore shifted to an attempt to understand the interaction between instructional control and contingency control, and also to understand the behavioural processes that determine what instructions are constructed and what control they acquire over behaviour.

Skinner's view of behaviour is most often characterised as a "molecular" view of behaviour; that is, each behaviour can be decomposed into atomistic parts or molecules. This view is inaccurate when one considers his complete description of behaviour as delineated in the 1981 article, Selection by Consequences and many other works. Skinner claims that a complete account of behaviour has involved an understanding of selection history at three levels: biology (the natural selection or phylogeny of the animal); behaviour (the reinforcement history or ontogeny of the behavioural repertoire of the animal); and for some species, culture (the cultural practices of the social group to which the animal belongs). This whole organism, with all those histories, then interacts with its environment. He often described even his own behaviour as a product of his phylogenetic history, his reinforcement history (which includes the learning of cultural practices) interacting with the environment at the moment. Molar behaviourists, such as Howard Rachlin argue that behaviour can not be understood by focusing on events in the moment. That is, they argue that a behaviour can be understood best in terms of the ultimate cause of history and that molecular behaviourist are committing a fallacy by inventing a fictitious proximal cause for behaviour. Molar behaviourists argue that standard molecular constructs such as "associative strength" are such fictitious proximal causes that simply take the place of molar variables such as rate of reinforcement. Thus, a molar behaviourist would define a behaviour such as loving someone as exhibiting a pattern of loving behaviour over time, there is no known proximal cause of loving behaviour, only a history of behaviours (of which the current behaviour might be an example of) that can be summarised as love. Molectular behaviourists use notions from Melioration theory, Negative power function discounting or additive versions of negative power function discounting.

Comparative Psychology

Comparative psychology viewed science less as a framework for explaining events but rather for predicting and controlling them. Adoption of the observational techniques that had proven useful in the study of the behaviour of animals was viewed as the best way to advance the scientific study of human beings. The external environment represented the main determining influence on behaviour and its manipulation held the most promise for improving the condition of human beings. The concept of a tabula rasa was revived.

A common tenet held that the complex phenomena of behaviour could be explained by and reduced to clear, distinct, reflexive units of behaviour. William James degined psychology as "the science of mental life." As a rigid application of the scientific method is only possible for learned behaviours, he felt that our studies should forrmost focus on studying those. Ivan P. Pavlov realised that behaviour changes when an organism comes to associate one stimulus with another. Specifically, a reflexive or automatic response transfers from one stimulus to another—*Classical Conditioning*.

Edward L. Thorndike Using a Puzzle Box, he demonstrated that behaviours that are followed by a positive outcome are often repeated, while those followed by a negative outcome or none at all are extinguished (i.e., Operant Conditioning). He postulated the Law of effect where behaviour changes when the results of an action proved of consequence (e.g. rewarding, noxious). John B. Watson stimulus-response theory of psychology, emotional reactions are learned in much the same way as other skills. B.F. Skinner contends that nearly all behaviour is shaped by complex patterns of reinforcement in a person's environment. Skinner Box: An animal placed inside the box is rewarded with a small bit of food each time it makes the desired response, such as pressing a lever or pecking a key. A device outside the box records the animal's responses.

Psychologists and scientists do not always agree on what should be considered Comparative Psychology. Taken in its most usual, broad sense, it refers to the study of the behaviour and mental life of animals other than human beings. Another possibility is that the emphasis should be placed on cross-species comparisons—including human to non-human animal comparisons. Some researchers however, feel that direct comparisons should not be the sole focus of comparative psychology and that intense focus on a single organism to understand its behaviour is just as desirable, if not more. Donald Dewsbury reviews the works of several psychologists and their definitions and concludes that the object of comparative psychology is to establish principles of generality focusing on both proximate and ultimate causation. It has been suggested that the term itself be discarded since it fails to be descriptive of the field but no appropriate replacement has been found. If looking for a precise definition, one may define comparative psychology as psychology concerned with the evolution (phylogenetic history and adaptive significance) and development (ontogenetic history and mechanism) of behaviour.

Using a comparative approach to behaviour allows one to evaluate the target behaviour from four different, complementary perspectives, developed by Niko Tinbergen. First, one may ask how pervasive the behaviour is across species? Meaning, how common is the behaviour in animals? Second, one may ask how the behaviour contributes to the lifetime reproductive success of the individuals demonstrating it? Meaning, does it result in those animals producing more offspring than animals not showing the behaviour? These two questions provide a theory for the ultimate cause of behaviour. Third, what mechanisms are involved in the behaviour? Meaning, what physiological, behavioural, and environmental components are necessary and sufficient for the generation of the behaviour? Fourth, a researcher may ask about the development of the behaviour within an individual. Meaning, what maturational, learning, social experiences must an individual undergo in order to demonstrate a behaviour? These latter two questions provide a theory for the proximate causes of behaviour. For more details see Tinbergen's four questions.

The earliest works on "the social organisation of ants" and "animal communication and psychology" were written by al-Jahiz, a 9th century Afro-Arab scholar who wrote many works on these subjects. The 11th century Arabic psychologist, Ibn al-Haytham (Alhazen), wrote the Treatise on the Influence of Melodies on the Souls of Animals, the early treatise dealing with the effects of music on animals. In the treatise, he demonstrates how a camel's pace could be hastened or retarded with the use of music, and shows other examples of how music can affect animal behaviour, experimenting with horses, birds and reptiles. Through to the 19th century, a majority of scholars in the Western world continued to believe that music was a distinctly human phenomenon, but experiments since then have vindicated Ibn al-Haytham's view that music does indeed have an effect on animals.

Charles Darwin was central in the development of comparative psychology; it is thought that psychology should be spoken in terms of "pre-" and "post-" Darwin because his contributions were so influential. Darwin's theory lead to several hypotheses, one being that the factors that set humans apart, such as higher mental, moral and spiritual faculties, could be accounted for by evolutionary principles. In response to the vehement opposition to Darwinism was the "anecdotal movement" led by George Romanes who set out to prove that animals possessed a "rudimentary human mind".

Near the end of the 19th century several scientists existed whose work was also very influential. Douglas Alexander Spalding, who was called the "first experimental biologist"worked mostly with birds—studying instinct, imprinting, and visual and auditory development. Jacques Loeb emphasised the importance of objectively studying behaviour, Sir John Lubbock is credited with first using mazes and puzzle devices to study learning and Lewis Henry Morgan is thought to be "the first ethologist in the sense in which we presently use the word".

Throughout the long history of comparative psychology, repeated attempts have been made to enforce a more disciplined approach, in which similar studies are carried out on animals of different species, and the results interpreted in terms of their different phylogenetic or ecological backgrounds. Behavioural ecology in the 1970's gave a more solid base of knowledge against which a true comparative psychology could develop. However, the broader use of the term "comparative psychology" is enshrined in the names of learned societies and academic journals, not to mention in the minds of psychologists of other specialisms, so it is never likely to disappear completely.

A persistent question with which comparative psychologists have been faced is the relative intelligence of different species of animal. Much effort has gone into explaining that this may not be a good question, but it will not go away. Indeed, some early attempts at a genuinely comparative psychology involved evaluating how well animals of different species could learn different tasks. These attempts floundered; in retrospect it can be seen that they were not sufficiently sophisticated, either in their analysis of the demands of different tasks, or in their choice of species to compare. More recent comparative work has been more successful, partly because it has drawn upon studies in ethology and behavioural ecology to make informed choices of species and tasks to compare.

A wide variety of species have been studied by comparative psychologists. However a small number have dominated the scene. Pavlov's early work used dogs, but although they have been the subject of occasional studies since then they have not figured prominently. Increasing interest in the study of abnormal animal behaviour has led to a return to the study of most kinds of domestic animal. Thorndike began his studies with cats, but American comparative psychologists quickly shifted to the more economical rat, which remained the almost invariable subject for the first half of the twentieth century and continues to be used. Skinner introduced the use of pigeons, and they continue to be important in some fields. There has always been interest in studying various species of primate; important contributions to social and developmental psychology were made by Harry F. Harlow's studies of maternal deprivation in rhesus monkeys. Interest in primate studies has increased with the rise in studies of animal cognition. Other animals thought to be intelligent have also been increasingly studied. Examples include various species of corvid, parrots — especially the African Gray Parrot — and dolphins.

Since the 1990s, comparative psychology has undergone a reversal in its fundamental approach. Instead of seeking principles in animal behaviour in order to explain human performance, comparative psychologists started taking principles that have been uncovered in the study of human cognition and testing them in animals of other species. This approach is referred to as the study of animal cognition. It has led to significant advances in our understanding of concept formation, memory, problem solving and other cognitive abilities in animals.

Today an animal's psychological constitution is recognised by veterinary surgeons as an important part of its living conditions in domestication or captivity. Common causes of disordered behaviour in captive or pet animals are lack of stimulation, inappropriate stimulation, or overstimulation. These conditions can lead to disorders, unpredictable and unwanted behaviour, and sometimes even physical symptoms and diseases. For example, rats that are exposed to loud music for a long period will ultimately develop unwanted behaviours that have been compared with human psychosis, like biting their owners.

The way dogs behave when understimulated is widely believed to depend on the breed as well as on the individual animal's character. For example, huskies have been known to completely ruin gardens and houses, if they are not allowed enough activity.Dogs are also prone to psychological damage if they are subjected to violence. If they are treated very badly, they may become dangerous. The systematic study of disordered animal behaviour draws on research in comparative psychology, including the early work on conditioning and instrumental learning, but also on ethological studies of natural behaviour. However, at least in the case of familiar domestic animals, it also draws on the accumulated experience of those who have worked closely with the animals.

Neuroscience

The last frontier of the biological sciences—their ultimate challenge—is to understand the biological basis of consciousness and the mental processes by which we perceive, act, learn, and remember.

The late 19th century witnessed the rise of modern neuroscience. Hypotheses that distinct brain functions were contained within specific regions of the brain emerged with evidence from patients with epileptic seizures, head trauma, and stroke. Extensive characterisations of neurons throughout the brain led Camillo Golgi and Santiago Ramón y Cajal to propose the neuron doctrine, i.e., that the function of the brain is a product of the electrical and chemical activity of individual neurons. Further support came from the discovery that muscles and neurons showed electrical excitability, which could be transfered onto neighboring cells.

Electrical properties are assessed either by recording from an electrode placed into the lumen of a cell (intracellular) or from its surrounding vicinity (extracellular). Intracellular recording techniques aim to either measure the current needed to maintain a constant voltage (voltage clamp) or measure the voltage needed to keep the current constant (current clamp) across the cell membrane. Using such techniques, Alan Hodgkin and Andrew Huxley identified the ionic mechanisms that underly the electrical properties of neurons in giant axons of squid. John Carew Eccles identified synapses as the sites of communication between neurons. Generations of neuroscientists had perceived the brain as a rather passive, reactive organ whose functional capacities were maintained by a sustained flow of sensory input.

The scope of neuroscience has now broadened to include any systematic scientific experimental and theoretical investigation of the central and peripheral nervous system of biological organisms. The empirical methodologies employed by neuroscientists have been enormously expanded, from biochemical and genetic analysis of dynamics of individual nerve cells and their molecular constituents to imaging representations of perceptual and motor tasks in the brain. Many recent theoretical advances in neuroscience have been aided by the use of computational modeling.

The scientific study of the nervous systems underwent a significant increase in the second half of the twentieth century, principally due to revolutions in molecular biology, electrophysiology, and computational neuroscience. It has become possible to understand, in much detail, the complex processes occurring within a single neuron. However, how networks of neurons produce intellectual behaviour, cognition, emotion, and physiological responses is still poorly understood.

The nervous system is composed of a network of neurons and other supportive cells (such as glial cells). Neurons form functional circuits, each responsible for specific tasks to the behaviours at the organism level. Thus, neuroscience can be studied at many different levels, ranging from molecular level to cellular level to systems level to cognitive level.

At the molecular level, the basic questions addressed in molecular· neuroscience include the mechanisms by which neurons express and respond to molecular signals and how axons form complex connectivity patterns. At this level, tools from molecular biology and genetics are used to understand how neurons develop and die, and how genetic changes affect biological functions. The morphology, molecular identity and physiological characteristics of neurons and how they relate to different types of behaviour are also of considerable interest. (The ways in which neurons and their connections are modified by experience are addressed at the physiological and cognitive levels.)

At the cellular level, the fundamental questions addressed in cellular neuroscience are the mechanisms of how neurons process signals physiologically and electrochemically. They address how signals are processed by the dendrites, somas and axons, and how neurotransmitters and electrical signals are used to process signals in a neuron.

At the systems level, the questions addressed in systems neuroscience include how the circuits are formed and used anatomically and physiologically to produce the physiological functions, such as reflexes, sensory integration, motor coordination, circadian rhythms, emotional responses, learning and memory. In other words, they address how these neural circuits function and the mechanisms through which behaviours are generated. For example, systems level analysis addresses questions concerning specific sensory and motor modalities: how does vision work? How do songbirds learn new songs and bats localise with ultrasound? The related field of neuroethology, in particular, addresses the complex question of how neural substrates underlies specific animal behaviour.

At the cognitive level, cognitive neuroscience addresses the questions of how psychological/cognitive functions are produced by the neural circuitry. The emergence of powerful new measurement techniques such as neuroimaging (e.g.,fMRI, PET, SPECT), electrophysiology and human genetic analysis combined with sophisticated experimental techniques from cognitive psychology allows neuroscientists and psychologists to address abstract questions such as how human cognition and emotion are mapped to specific neural circuitries.

Neuroscience is also beginning to become allied with social sciences, and burgeoning interdisciplinary fields of neuroeconomics, decision theory, social neuroscience are starting to address some of the most complex questions involving interactions of brain with environment.

Neuroscience generally includes all scientific studies involving the nervous system. Psychology, as the scientific study of mental processes, may be considered a sub-field of neuroscience, although some mind/body theorists argue that the definition goes the other way — that psychology is a study of mental processes that can be modeled by many other abstract principles and theories, such as behaviourism and traditional cognitive psychology, that are independent of the underlying neural processes. The term neurobiology is sometimes used interchangeably with neuroscience, though the former refers to the biology of nervous system, whereas the latter refers to science of mental functions that form the foundation of the constituent neural circuitries. In Principles of Neural Science, nobel laureate Eric Kandel contends that cognitive psychology is one of the pillar disciplines for understanding the brain in neuroscience.

Neurology and Psychiatry are medical specialties that specifically address the diseases of the nervous system. These terms also refer to clinical disciplines involving diagnosis and treatment of these diseases. Neurology deals with diseases of the central and peripheral nervous systems such as amyotrophic lateral sclerosis (ALS) and stroke, while psychiatry focuses on behavioural, cognitive, and emotional disorders. The boundaries between the two have been blurring recently and physicians who specialise in either generally receive training in both. Both neurology and psychiatry are heavily influenced by basic research in neuroscience.

Evidence of trepanation, the surgical practice of either drilling or scraping a hole into the skull with the aim of curing headaches or mental disorders or relieving cranial pressure, being performed on patients dates back to Neolithic times and has been found in various cultures throughout the world. Manuscripts dating back to 5000BC indicated that the Egyptians had some knowledge about symptoms of brain damage.

Early views on the function of the brain regarded it to be a "cranial stuffing" of sorts. In Egypt, from the late Middle Kingdom onwards, the brain was regularly removed in preparation for mummification. It was believed at the time that the heart was the seat of intelligence. According to Herodotus, during the first step of mummification: 'The most perfect practice is to extract as much of the brain as possible with an iron hook, and what the hook cannot reach is mixed with drugs.'

The view that the heart was the source of consciousness was not challenged until the time of Hippocrates. He believed that the brain was not only involved with sensation, since most specialised organs (e.g., eyes, ears, tongue) are located in the head near the brain, but was also the seat of intelligence. Aristotle, however, believed that the heart was the center of intelligence and that the brain served to cool the blood. This view was generally accepted until the Roman physician Galen, a follower of Hippocrates and physician to Roman gladiators, observed that his patients lost their mental faculties when they had sustained damage to their brains.

In al-Andalus, Abulcasis, the father of modern surgery, developed material and technical designs which are still used in neurosurgery. Averroes suggested the existence of Parkinson's disease and attributed photoreceptor properties to the retina. Avenzoar described meningitis, intracranial thrombophlebitis, mediastinal tumours and made contributions to modern neuropharmacology. Maimonides wrote about neuropsychiatric disorders and described rabies and belladonna intoxication.Elsewhere in medieval Europe, Vesalius (1514-1564) and René Descartes (1596-1650) also made several contributions to neuroscience.

Studies of the brain became more sophisticated after the invention of the microscope and the development of a staining procedure by Camillo Golgi during the late 1890s that used a silver chromate salt to reveal the intricate structures of single neurons. His technique was used by Santiago Ramón y Cajal and led to the formation of the neuron doctrine, the hypothesis that the functional unit of the brain is the neuron. Golgi and Ramón y Cajal shared the Nobel Prize in Physiology or Medicine in 1906 for their extensive observations, descriptions and categorisations of neurons throughout the brain. The hypotheses of the neuron doctrine were supported by experiments following Galvani's pioneering work in the electrical excitability of muscles and neurons. In the late 19th century, DuBois-Reymond, Müller, and von Helmholtz showed neurons were electrically excitable and that their activity predictably affected the electrical state of adjacent neurons.

In parallel with this research, work with brain-damaged patients by Paul Broca suggested that certain regions of the brain were responsible for certain functions. At the time Broca's findings were seen as a confirmation of Franz Joseph Gall's theory that language was localised and certain psychological functions were localised in the cerebral cortex. The localisation of function hypothesis was supported by observations of epileptic patients conducted by John Hughlings Jackson, who correctly deduced the organisation of motor cortex by watching the progression of seizures through the body. Wernicke further developed the theory of the specialisation of specific brain structures in language comprehension and production. Modern research still uses the Brodmann cytoarchitectonic (referring to study of cell structure) anatomical definitions from this era in continuing to show that distinct areas of the cortex are activated in the execution of specific tasks.

Ethology

Ethology is the scientific study of animal behaviour. Although many naturalists have studied aspects of animal behaviour through the centuries, the modern discipline of ethology is usually considered to have arisen with the work in the 1930s of Dutch biologist Nikolaas Tinbergen and Austrian biologist Konrad Lorenz, joint winners of the 1973 Nobel Prize in medicine. Ethology is a combination of laboratory and field science, with strong ties to certain other disciplines — e.g., neuroanatomy, ecology, evolution. Ethologists are typically interested in a behavioural process rather than in a particular animal group and often study one type of behaviour (e.g. aggression) in a number of unrelated animals.

The desire to understand the animal world has made ethology a rapidly growing field, and since the turn of the 21st century, many prior understandings related to diverse fields such as animal communication, personal symbolic name use, animal emotions, animal culture and learning, and even sexual conduct, long thought to be well understood, have been revolutionised, as have new fields such as neuroethology.

Comparative psychology also studies animal behaviour, but, as opposed to ethology, construes its study as a branch of psychology rather than as one of biology. Historically, where comparative psychology sees the study of animal behaviour in the context of what is known about human psychology, ethology sees the study of animal behaviour in the context of what is known about animal anatomy, physiology, neurobiology, and phylogenetic history. This distinction is not representative of the current state of the field. Furthermore, early comparative psychologists concentrated on the study of learning and tended to look at behaviour in artificial situations, whereas early ethologists concentrated on behaviour in natural situations, tending to describe it as instinctive.

The two approaches are complementary rather than competitive, but they do lead to different perspectives and sometimes to conflicts of opinion about matters of substance. In addition, for most of the twentieth century, comparative psychology developed most strongly in North America, while ethology was stronger in Europe, and this led to different emphases as well as somewhat differing philosophical underpinnings in the two disciplines. A practical difference is that early comparative psychologists concentrated on gaining extensive knowledge of the behaviour of very few species, while ethologists were more interested in gaining knowledge of behaviour in a wide range of species in order to be able to make principled comparisons across taxonomic groups. Ethologists have made much more use of a truly comparative method than comparative psychologists have. Despite the historical divergence, most ethologists (as opposed to behavioural ecologists), at least in North America, teach in psychology departments. It is a strong belief among scientists that the mechanisms on which behavioural processes are based are the same that rule the evolution of the living species: there is therefore a strong connection between these two fields.

Until the 18th century, the most common theory among scientists was still the Scala Naturae proposed by Aristotle: according to this theory, the living beings were classified on an ideal pyramid in which the simplest animals were occupying the lower floors, and then complexity would raise progressively until the top, which was occupied by the human beings. There was also an avant-garde group of biologists who were refusing the Aristotelian theory for a more anthropocentric one, according to which all living beings were created by God to serve mankind, and would behave accordingly. A well-radicated opinion in the common sense of the time in the Western world was that animal species were eternal and immutable, created with a specific purpose, as this seemed the only possible explanation for the incredible variety of the living beings and their surprising adaptation to their habitat.

The first biologist elaborating a complex evolution theory was Jean-Baptiste Lamarck (1744-1829). His theory was substantially made of two statements: the first is that animal organs and behaviour can change according to the way they are being used, and that those characteristics are capable of being transmitted from one generation to the next (well-known is the example of the giraffe whose neck becomes longer while trying to reach the upper leaves of a tree). The second affirmation is that each and every living organism, human beings included, tends to reach a greater level of perfection. At the time of his journey for the Galapagos Islands, Charles Darwin was well aware of Lamarck's theories and was deeply influenced by them.

Because ethology is understood as a branch of biology, ethologists have been particularly concerned with the evolution of behaviour and the understanding of behaviour in terms of the theory of natural selection. In one sense, the first modern ethologist was Charles Darwin, whose book, The Expression of the Emotions in Man and Animals, has influenced many ethologists. He pursued his interest in behaviour by encouraging his protégé George Romanes, who investigated animal learning and intelligence using an anthropomorphic method, anecdotal cognitivism, that did not gain scientific support.

Other early ethologists, such as Oskar Heinroth and Julian Huxley, instead concentrated on behaviours that can be called instinctive, or natural, in that they occur in all members of a species under specified circumstances. Their first step in studying the behaviour of a new species was to construct an ethogram (a description of the main types of natural behaviour with their frequencies of occurrence). This approach provided an objective, cumulative base of data about behaviour, which subsequent researchers could check and build upon.

An important step, associated with the name of Konrad Lorenz though probably due more to his teacher, Oskar Heinroth, was the identification of fixed action patterns (FAPs). Lorenz popularised FAPs as instinctive responses that would occur reliably in the presence of identifiable stimuli (called sign stimuli or releasing stimuli). These FAPs could then be compared across species, and the similarities and differences between behaviour could be easily compared with the similarities and differences in morphology. An important and much quoted study of the Anatidae (ducks and geese) by Heinroth used this technique. The ethologists noted that the stimuli that released FAPs were commonly features of the appearance or behaviour of other members of their own species, and they were able to show how important forms of animal communication could be mediated by a few simple FAPs. The most sophisticated investigation of this kind was the study by Karl von Frisch of the so-called "dance language" underlying bee communication. Lorenz developed an interesting theory of the evolution of animal communication based on his observations of the nature of fixed action patterns and the circumstances in which animals emit them.

The Merriam-Webster dictionary defines instinct as a largely inheritable and unalterable tendency of an organism to make a complex and specific response to environmental stimuli without involving reason. In ethology, by instinct is meant a series of predictable ways and behavioural footsteps which go under fixed action patterns. Such schemes are only acted when a precise stimulating signal is present. When such signals act as communication among members of the same species, they go under the name of releasers. Notable examples of releasers are, in many bird species, the beak movements by the newborns, which stimulates the mother's regurgitating process to feed her offspring. Another well known case is the classic experiments by Tinbergen and Lorenz on the Graylag Goose. Like similar waterfowl, it will roll a displaced egg near its nest back to the others with its beak.

The sight of the displaced egg triggers this mechanism. If the egg is taken away, the animal continues with the behaviour, pulling its head back as if an imaginary egg is still being maneuvered by the underside of its beak. However, it will also attempt to move other egg shaped objects, such as a golf ball, door knob, or even an egg too large to have possibly been laid by the goose itself (a supernormal stimulus). As made obvious by this last example, however, a behaviour only made of fixed action patterns would be particularly rigid and inefficient, reducing the probabilities of survival and reproduction. The learning process has therefore great importance, as the ability to change the individual's responses based on its experience. It can be said that the more the brain is complex and the life of the individual long, the more its behaviour will be "intelligent" (in the sense of guided by experience rather than rigid FAPs).

The learning process may take place in many ways, one of the most elementary is habituation. This process consists in ignoring persistent or useless stimuli. An example of learning by habituation is the one observed in squirrels: when one of them feels in danger, the others hear its signal and go to the nearest repair. However, if the signal comes from an individual who has performed a big number of false alarms, his signal will be ignored.

Another common way of learning is by association, where a stimuli is, based on the experience, linked to another one which may not have anything to do with the first one. The first studies of associative learning were made by Russian physiologist Ivan Pavlov. An example of associative behaviour is observed when a common goldfish goes close to the water surface whenever a human is going to feed it, or the excitement of a dog whenever it sees a collar as a prelude for a walk. The associative learning process is linked to the necessity of developing discriminatory capacities, that is, the faculty of making meaningful choices. Being able to discriminate the members of your own species is of fundamental importance for the reproductive success. Such discrimination can be based on a number of factors in many species including birds, however, this important type of learning only takes place in a very limited period of time. This kind of learning is called imprinting.

A second important finding of Lorenz concerned the early learning of young nidifugous birds, a process he called imprinting. Lorenz observed that the young of birds such as geese and chickens spontaneously followed their mothers from almost the first day after they were hatched, and he discovered that this response could be imitated by an arbitrary stimulus if the eggs were incubated artificially and the stimulus was presented during a critical period (a less temporally constrained period is called a sensitive period) that continued for a few days after hatching.

Finally, imitation is often a big part of the learning process. A well-documented example of imitative learning is that of macaques in Hachijojima island, Japan. These primates used to live in the inland forest until the 60s, when a group of researchers started giving them some potatoes on the beach: soon they started venturing onto the beach, picking the potatoes from the sand, and cleaning and eating them. About one year later, an individual was observed bringing a potato to the sea, putting it into the water with one hand, and cleaning it with the other. Her behaviour was soon imitated by the individuals living in contact with her; when they gave birth, they taught this practice to their young.

The individual reproduction is with no doubt the most important phase in the proliferation of the species: for this reason, we can often observe complex mating rituals, which can reach a high level of complexity even if they are often regarded as fixed action patterns (FAPs). The Stickleback's complex mating ritual was studied by Niko Tinbergen and is regarded as a notable example of a FAP. Often in social life, males are fighting for the right of reproducing themselves as well as social supremacy. Such behaviours are common among mammals.

A common example of fight for social and sexual supremacy is the so-called pecking order among poultry. A pecking order is established every time a group of poultry co-lives for a certain amount of time. In each of these groups, a chicken is dominating among the others and can peck before anyone else without being pecked. A second chicken can peck all the others but the first, and so on. The chicken in the higher levels can be easily distinguished for their well-cured aspect, as opposed to the ones in the lower levels. During the period in which the pecking order is establishing, frequent and violent fights can happen, but once it is established it is only broken when other individuals are entering the group, in which case the pecking order has to be established from scratch.

Several animal species, including humans, tend to live in groups. Group size is a major aspect of their social environment. Social life is probably a complex and effective survival strategy. It may be regarded as a sort of symbiosis among individuals of the same species: a society is composed of a group of individuals belonging to the same species living within well-defined rules on food management, role assignments and reciprocal dependence.

The situation is, actually much more complex than it looks. When biologists interested in evolution theory first started examining social behaviour, some apparently unanswerable questions came up. How could, for instance, the birth of sterile casts, like in bees, be explained through an evolving mechanism which emphasizes the reproductive success of as many individuals as possible? Why, among animals living in small groups like squirrels, would an individual risk its own life to save the rest of the group? These behaviours may be examples of altruism. Of course, not all behaviours are altruistic, as shown in the table below. Notably, revengeful behaviour is claimed to have been observed exclusively in Homo sapiens.

Table 1. Classification of social behaviours

Type of behaviour	*Effect on the donor*	*Effect on the receiver*
Egoistic	Increases fitness	Decreases fitness
Cooperative	Increases fitness	Increases fitness
Altruistic	Decreases fitness	Increases fitness
Revengeful	Decreases fitness	Decreases fitness

The existence of egoism through natural selection doesn't pose any question to the evolution theory and is, on the contrary, fully justified by it, as well as for the cooperative behaviour. It is much harder to understand the mechanism through which the altruistic behaviour initially developed.

Behavioural Genetics

Behavioural Genetics explores the heritability of behavioural traits. More recently, this has given rise to a largely quantitative emphasis on using molecular genetic techniques in order to isolate individual genes that play a role in behaviour, neural and behavioural pathologies, or individual human charcteristics.

Each inherited feature of an organism is determined by a section of genetic code—a gene which represents the smallest hereditary unit. The genome refers to one individual's entire set of hereditary information and includes genes as well as non-coding sequences of DNA. The genomes for any two individuals, except for those of identical twins or cloned animals, are unique. Such differences (i.e., polymorphism) produce different versions of the same genes, called alleles. Despite these differences in genetic makeup, we also find major commonalities across much of the code shared by all forms of life. The specific genetic makeup of an individual is called its genotype. A trait refers to a specific characteristic or feature of an organism. The term phenotype represents the specific manifestation of a trait or its physical appearance and constitution and refers to the actual state of that trait. The individual's phenotype emerges as as interation of genetic information with environmental conditions.

Currently, the largest branch of human behavioural genetics is psychiatric genetics which studies phenotypes such as schizophrenia, bipolar disorder, and alcoholism. Recent trends in behaviour genetics have indicated an additional focus toward researching the inheritance of human characteristics typically studied in developmental psychology. For instance, a major focus in developmental psychology has been to characterise the influence of parenting styles on children. However, in most studies, genes are a confounding variable. Because children share half of their genes with each parent, any observed effects of parenting styles could be effects of having many of the same genes as a parent (e.g. harsh aggressive parenting styles have been found to correlate with similar aggressive child characteristics: is it the parenting or the genes?). Thus, behaviour genetics research is currently undertaking to distinguish the effects of the family environment from the effects of genes. This branch of behaviour genetics research is becoming more closely associated with mainstream developmental psychology and the sub-field of developmental psychopathology as it shifts its focus to the heritability of such factors as emotional self-control, attachment, social functioning, aggressiveness, etc.

Several academic bodies exist to support behaviour genetic research, including the International Behavioural and Neural Genetics Society, Behaviour Genetics Association, the International Society for Psychiatric Genetics, and the International Society for Twin Studies. Behaviour genetic work features prominently in several more general societies, for instance the International Behavioural Neuroscience Society.

Human behavioural geneticists use several designs to answer questions about the nature and mechanisms of genetic influences on behaviour. All of these designs are unified by being based around human relationships which disentangle genetic and environmental relatedness. So, for instance, some researchers study adopted twins: the adoption study. In this case the adoption disentangles the genetic relatedness of the twins (either 50% or 100%) from their family environments. Likewise the classic twin study contrasts the differences between identical twins and fraternal twins within a family compared to differences observed between families. This core design can be extended: the so-called "extended twin study" which adds additional family members, increasing power and allowing new genetic and environmental relationships to be studied. Excellent examples of this model are the Virginia 20,000 and the QIMR twin studies.

Also possible are the "children of twins" design (holding maternal genetic contributions equal across children with paternal genetics and family environments; and the "virtual twins" design - unrelated children adopted into a family who are very close or identical in age to biological children or other adopted children in the family. While the classical twin study has been criticised they continue to be of high utility. There are several dozen major studies ongoing, in countries as diverse as the USA, UK, Germany, France, The Netherlands, and Australia, and the method is used widely in fields as diverse as dental caries, BMI, aging, substance abuse, sexuality, cognitive abilities, personality, values, and a wide range of psychiatry disorders. This is broad utility is reflected in several thousands of peer-review papers, and several dedicated societies and journals.

Behavioural Ecology

Behavioural ecology is the study of the ecological and evolutionary basis for animal behaviour, and the roles of behaviour in enabling an animal to adapt to its environment (both intrinsic and extrinsic). Behavioural ecology emerged from ethology after Niko Tinbergen (a seminal figure in the study of animal behaviour), outlined the four causes of behaviour.

The two causes that contribute to ultimate causation are phylogenetic constraints and adaptive significance. Phylogenetic constraints are factors that might stop certain lineages developing certain behavioural or morphological traits. Hence, it is no coincidence that generally birds are able to fly and mammals cannot. The evolutionary history of these lineages have made it profitable for birds to fly and for mammals to walk. Adaptive significance is akin to asking what a trait is good for in an evolutionary context. Therefore, the adaptive significance of flight in birds might have enabled avian ancestors to escape from predators. However, it is not sufficient to apply these explanations where they seem convenient. These have been labeled Just So Stories by some biologists after Rudyard Kipling's tales for children about how animals came to be the way they are. To be rigorous, one must suppose a hypothesis and then test it scientifically. Hence, for avian flight, one can suppose that when birds are not at risk of being eaten, they might lose the ability to fly.

Proximate causation is also divided into two factors which are ontogenetic and mechanistic. Ontogenetic factors are the entire sum of experience throughout the lifetime of an individual from embryo to death. Hence, factors included are learning the genetic factors giving rise to behaviour in individuals. Mechanistic factors, as the name implies, are the processes of the body that give rise to behaviour such as the effects of hormones on behaviour and neuronal basis of behaviour.

Behavioural ecology, along with other areas of evolutionary biology, has incorporated a number of techniques which have been borrowed from optimisation theory. Optimisation is a concept that stipulates strategies that offer the highest return to an animal given all the different factors and constraints facing the animal. One of the simplest ways to arrive at an optimal solution is to do a cost/benefit analysis. By considering the advantages of a behaviour and the costs of a behaviour, it can be seen that if the costs outweigh the benefits then a behaviour will not evolve and vice versa. This is also where the concept of the trade-off becomes important. This is because it rarely pays an animal to invest maximally in any one behaviour.

For example, the amount of time an ectothermic animal such as a lizard spends foraging is constrained by its body temperature. The digestive efficiency of the lizard also increases with increases in body temperature. Lizards increase their body temperature by basking in the sun. However, the time spent basking decreases the amount of time available for foraging. Basking also increases the risk of being discovered by a predator. Therefore, the optimal basking time is the outcome of the time necessary to sufficiently warm itself to carry out its activities such as foraging. This example shows how foraging is constrained by the need to bask (intrinsic constraint) and predation pressure (extrinsic constraint).

Ultimately, behaviour is subject to natural selection just as with any other trait. Therefore animals that employ optimal behavioural strategies specific to their environment will generally leave greater numbers of offspring than their suboptimal conspecifics. Animals that leave a greater number of offspring than others of their own species are said to have greater fitness. However, environments change over time. What might be good behaviour today might not be the best behaviour in 10,000 years time or even 10 years time. The behaviour of animals has and will continue to change in response to the environment. Behavioural ecology is one of the best ways to study these changes. As geneticist Theodosius Dobzhansky famously wrote, "nothing in biology makes sense except in the light of evolution."

Another driving force in the evolution of animal behaviour is the concept of an evolutionarily stable strategy (or ESS), a term derived from economic game theory which became prominent after John Maynard Smith's 1982 book, Evolution and the Theory of Games. However, the concept can be traced back (as with most evolutionary ideas) to W.D. Hamilton, R.A. Fisher and Charles Darwin. In short, evolutionary game theory asserts that only strategies that, when common in the population, cannot be "invaded" by any alternative rare (mutant) strategy will be ESSs, and thus maintained in the population. Therefore, animal behaviour can be said to be governed not only by what is optimal, but also by what other strategies are found in the population. Furthermore, the relative frequencies of each strategy can influence the fitness of the other strategies in the population (frequency dependence). It is important to consider that evolution is not only driven by the physical environment, but also the interactions between other individuals.

Sociobiology

Sociobiology is a neo-Darwinian synthesis of scientific disciplines that attempts to explain social behaviour in all species by considering the evolutionary advantages the behaviours may have. It is often considered a branch of biology and sociology, but also draws from ethology, anthropology, evolution, zoology, archaeology, population genetics and other disciplines. Within the study of human societies, sociobiology is closely related to the fields of human behavioural ecology and evolutionary psychology.

Sociobiology investigates social behaviours, such as mating patterns, territorial fights, pack hunting, and the hive society of social insects. Just as selection pressure led to animals evolving useful ways of interacting with the natural environment, it led to the genetic evolution of advantageous social behaviour. Applied to non-humans, sociobiology is uncontroversial.

Sociobiology has become one of the greatest scientific controversies of the late 20th and early 21st centuries, especially in the context of explaining human behaviour. Criticism, most notably made by Richard Lewontin and Stephen Jay Gould, centers on sociobiology's contention that genes play a central role in human behaviour and that variation in traits such as aggressiveness can be explained by variation in peoples' biology and is not necessarily a product of the person's social environment. Many sociobiologists, however, cite a complex relationship between nature and nurture. In response to the controversy, anthropologist John Tooby and psychologist Leda Cosmides launched evolutionary psychology as a branch of sociobiology made less controversial by avoiding questions of human biodiversity.

Sociobiology is based on the idea that some behaviours (both social and individual) are at least partly inherited and can be affected by natural selection. It starts with the idea that these behaviours have evolved over time, similar to the way that physical traits are thought to have evolved. Therefore, it predicts that animals will act in ways that have proven to be evolutionarily successful over time, which can among other things result in the formation of complex social processes that have proven to be conducive to evolutionary fitness.

The discipline seeks to explain behaviour as a product of natural selection; thus behaviour is seen as an effort to preserve one's genes in the population. Inherent in sociobiological reasoning is the idea that certain genes or gene combinations that influence particular behavioural traits can be "passed down" from generation to generation.

For example, newly dominant male lions often will kill cubs in the pride that were not sired by them. This behaviour is adaptive in evolutionary terms because killing the cubs eliminates competition for their own offspring and causes the nursing females to come into heat faster, thus allowing more of his genes to enter into the population. Sociobiologists would view this instinctual cub-killing behaviour as being "passed down" through the genes of successfully reproducing male lions, whereas non-killing behaviour may have "died out" as those lions were less successful in reproducing.

Genetic mouse mutants have now been harnessed to illustrate the power that genes exert on behaviour. For example, the transcription factor FEV (aka Pet1) has been shown, through its role in maintaining the serotonergic system in the brain, to be required for normal aggressive and anxiety-like behaviour. Thus, when FEV is genetically deleted from the mouse genome, male mice will instantly attack other males, whereas their wild-type counterparts take significantly longer to initiate violent behaviour. In addition, FEV has been shown to be required for correct maternal behaviour in mice, such that their offspring do not survive unless cross-fostered to other wild-type female mice. A genetic basis for instinctive behavioural traits among non-human species, such as in the above example, is commonly accepted among many biologists; however, attempting to use a genetic basis to explain complex behaviours in human societies has remained extremely controversial.

According to the OED, John Paul Scott coined the word "sociobiology" at a 1946 conference on genetics and social behaviour, and became widely used after it was popularised by Edward O. Wilson in his 1975 book, Sociobiology: The New Synthesis. However, the influence of evolution on behaviour has been of interest to biologists and philosophers since soon after the discovery of the evolution itself. Peter Kropotkin's Mutual Aid: A Factor of Evolution, written in the early 1890s, is a popular example. Antecedents of modern sociobiological thinking can be traced to the 1960s and the work of such biologists as Robert Trivers and William D. Hamilton.

Nonetheless, it was Wilson's book that pioneered and popularised the attempt to explain the evolutionary mechanics behind social behaviours such as altruism, aggression, and nurturence, primarily in ants (Wilson's own research specialty) but also in other animals. The final chapter of the book is devoted to sociobiological explanations of human behaviour, and Wilson later wrote a Pulitzer Prize winning book, On Human Nature, that addressed human behaviour specifically.

Sociobiologists believe that human behaviour, as well as nonhuman animal behaviour, can be partly explained as the outcome of natural selection. They contend that in order fully to understand behaviour, it must be analysed in terms of evolutionary considerations.

Natural selection is fundamental to evolutionary theory. Variants of hereditary traits which increase an organism's ability to survive and reproduce will be more greatly represented in subsequent generations, i.e., they will be "selected for". Thus, inherited behavioural mechanisms that allowed an organism a greater chance of surviving and/or reproducing in the past are more likely to survive in present organisms. That inherited adaptive behaviours are present in nonhuman animal species has been multiply demonstrated by biologists, and it has become a foundation of evolutionary biology. However, there is continued resistance by some researchers over the application of evolutionary models to humans, particularly from within the social sciences, where culture has long been assumed to be the predominant driver of behaviour.

Sociobiology is based upon two fundamental premises:

— Certain behavioural traits are inherited,
— Inherited behavioural traits have been honed by natural selection. Therefore, these traits were probably "adaptive" in the species` evolutionarily evolved environment.

Sociobiology uses Nikolaas Tinbergen's four categories of questions and explanations of animal behaviour. Two categories are at the species level; two, at the individual level. The species-level categories (often called "ultimate explanations") are:

— the function (i.e., adaptation) that a behaviour serves and
— the evolutionary process (i.e., phylogeny) that resulted in this functionality.

The individual-level categories are

— the development of the individual (i.e., ontogeny) and
— the proximate mechanism (e.g., brain anatomy and hormones).

Sociobiologists are interested in how behaviour can be explained logically as a result of selective pressures in the history of a species. Thus, they are often interested in instinctive, or intuitive behaviour, and in explaining the similarities, rather than the differences, between cultures. For example, mothers within many species of mammals – including humans – are very protective of their offspring. Sociobiologists reason that this protective behaviour likely evolved over time because it helped those individuals which had the characteristic to survive and reproduce. Over time, individuals who exhibited such protective behaviours would have had more surviving offspring than did those who did not display such behaviours, such that this parental protection would increase in frequency in the population. In this way, the social behaviour is believed to have evolved in a fashion similar to other types of nonbehavioural adaptations, such as (for example) fur or the sense of smell.

Individual genetic advantage often fails to explain certain social behaviours as a result of gene-centred selection, and evolution may also act upon groups. The mechanisms responsible for group selection employ paradigms and population statistics borrowed from game theory. E.O. Wilson argued that altruistic individuals must reproduce their own altruistic genetic traits for altruism to survive. When altruists lavish their resources on non-altruists at the expense of their own kind, the altruists tend to die out and the others tend to grow. In other words, altruism is more likely to survive if altruists practice the ethic that "charity begins at home."

Within sociobiology, a social behaviour is first explained as a sociobiological hypothesis by finding an evolutionarily stable strategy that matches the observed behaviour. Stability of a strategy can be difficult to prove, but usually, a well-formed strategy will predict gene frequencies. The hypothesis can be supported by establishing a correlation between the gene frequencies predicted by the strategy, and those expressed in a population. Measurement of genes and gene-frequencies can be problematic, however, because a simple statistical correlation can be open to charges of circularity (Circularity can occur if the measurement of gene frequency indirectly uses the same measurements that describe the strategy).

Altruism between social insects and littermates has been explained in such a way. Altruistic behaviour in some animals has been correlated to the degree of genome shared between altruistic individuals. A quantitative description of infanticide by male harem-mating animals when the alpha male is displaced. Female infanticide and fetal resorption in rodents are active areas of study. In general, females with more bearing opportunities may value offspring less. Also, females may arrange bearing opportunities to maximise the food and protection from mates.

An important concept in sociobiology is that temperamental traits within a gene pool and between gene pools exist in an ecological balance. Just as an expansion of a sheep population might encourage the expansion of a wolf population, an expansion of altruistic traits within a gene pool may also encourage the expansion of individuals with dependent traits.

Sociobiology is often mistakenly associated with arguments over the "genetic" basis of intelligence. While sociobiology is predicated on the observation that genes do affect behaviour, it is perfectly consistent to be a sociobiologist while arguing that measured IQ variations between individuals reflect mainly cultural or economic rather than genetic factors. However, many critics point out that the usefulness of sociobiology as an explanatory tool breaks down once a trait is so variable as to no longer be exposed to selective pressures. In order to explain aspects of human intelligence as the outcome of selective pressures, it must be demonstrated that those aspects are inherited, or genetic, but this does not necessarily imply differences among individuals: a common genetic inheritance could be shared by *all* humans, just as the genes responsible for number of limbs are shared by all individuals.

Researchers performing twin studies have argued that differences between people on behavioural traits such as creativity, extroversion and aggressiveness are between 45% to 75% due to genetic differences, and intelligence is said by some to be about 80% genetic after one matures (discussed at Intelligence quotient#Genetics vs environment). However, critics (such as the evolutionary geneticist R. C Lewontin) have highlighted serious flaws in twin studies, such as the inability of researchers to separate environmental, genetic, and dialectic effects on twins, and twin studies as a tool for determining the heritability of behavioural traits in humans have been largely abandoned.

Evolutionary Psychology

Evolutionary psychology (EP) is an approach to the entire discipline that views human nature as a universal set of evolved psychological adaptations to recurring problems in the ancestral environment. Proponents of EP suggest that it seeks to heal a fundamental division at the very heart of science — that between the soft human social sciences and the hard natural sciences, and that the fact that human beings are living organisms demands that psychology be understood as a branch of biology. Anthropologist John Tooby and psychologist Leda Cosmides note:

> "Evolutionary psychology is the long-forestalled scientific attempt to assemble out of the disjointed, fragmentary, and mutually contradictory human disciplines a single, logically integrated research framework for the psychological, social, and behavioural sciences—a framework that not only incorporates the evolutionary sciences on a full and equal basis, but that systematically works out all of the revisions in existing belief and research practice that such a synthesis requires."

Just as human physiology and evolutionary physiology have worked to identify physical adaptations of the body that represent "human physiological nature," the purpose of evolutionary psychology is to identify evolved emotional and cognitive adaptations that represent "human psychological nature." EP is, to quote Steven Pinker, "not a single theory but a large set of hypotheses" and a term which "has also come to refer to a particular way of applying evolutionary theory to the mind, with an emphasis on adaptation, gene-level selection, and modularity." EP proposes that the human brain comprises many functional mechanisms, called psychological adaptations or evolved cognitive mechanisms or cognitive modules designed by the process of natural selection. Examples include language acquisition modules, incest avoidance mechanisms, cheater detection mechanisms, intelligence and sex-specific mating preferences, foraging mechanisms, alliance-tracking mechanisms, agent detection mechanisms, and others.

EP has roots in cognitive psychology and evolutionary biology. It also draws on behavioural ecology, artificial intelligence, genetics, ethology, anthropology, archaeology, biology, and zoology. EP is closely linked to sociobiology, but there are key differences between them including the emphasis on domain-specific rather than domain-general mechanisms, the relevance of measures of current fitness, the importance of mismatch theory, and psychology rather than behaviour. Many evolutionary psychologists, however, argue that the mind consists of both domain-specific and domain-general mechanisms, especially evolutionary developmental psychologists. Most sociobiological research is now conducted in the field of behavioural ecology.

The term evolutionary psychology was probably coined by American biologist Michael Ghiselin in a 1973 article published in the journal Science. Jerome Barkow, Leda Cosmides and John Tooby popularised the term "evolutionary psychology" in their highly influential 1992 book The Adapted Mind: Evolutionary Psychology and The Generation of Culture. EP has been applied to the study of many fields, including economics, aggression, law, psychiatry, politics, literature, and sex.

The species-level categories (often called "ultimate explanations") are

- the function (i.e.,adaptation) that a behaviour serves and
- the evolutionary process (i.e., phylogeny) that resulted in the adaptation (functionality).

The individual-level categories are:

- the development of the individual (i.e., ontogeny) and
- the proximate mechanism (e.g., brain anatomy and hormones).

Evolutionary psychology is a hybrid discipline that draws insights from modern evolutionary theory, biology, cognitive psychology, anthropology, economics, computer science, and paleoarchaeology. The discipline rests on a foundation of core premises. According to evolutionary psychologist David Buss, these include:

- Manifest behaviour depends on underlying psychological mechanisms, information processing devices housed in the brain, in conjunction with the external and internal inputs that trigger their activation.
- Evolution by selection is the only known causal process capable of creating such complex organic mechanisms.
- Evolved psychological mechanisms are functionally specialised to solve adaptive problems that recurred for humans over deep evolutionary time.

— Selection designed the information processing of many evolved psychological mechanisms to be adaptively influenced by specific classes of information from the environment.

Human psychology consists of a large number of functionally specialised evolved mechanisms, each sensitive to particular forms of contextual input, that get combined, coordinated, and integrated with each other to produce manifest behaviour.

Similarly, pioneers of the field Leda Cosmides and John Tooby consider five principles to be the foundation of evolutionary psychology:

— The brain is a physical system. It functions as a computer with circuits that have evolved to generate behaviour that is appropriate to environmental circumstances.

— Neural circuits were designed by natural selection to solve problems that human ancestors faced while evolving into Homo sapiens.

— Consciousness is a small portion of the contents and processes of the mind; conscious experience can mislead individuals to believe their thoughts are simpler than they actually are. Most problems experienced as easy to solve are very difficult to solve and are driven and supported by very complicated neural circuitry

— Evolutionary psychology is founded on the computational theory of mind, the theory that the mind, our "inner world," is the action of complex neural structures in the brain. For example, when a child shrinks in fear from a spider, the child's brain has attended to the spider, computed that it's a potential threat, and initiated a fear response.

— Evolutionary psychology is rooted in evolutionary theory. It is sometimes seen not simply as a sub-discipline of psychology but as a way in which evolutionary theory can be used as a meta-theoretical framework within which to examine the entire field of psychology. A few biologists challenge the basic premises of evolutionary psychology.

Natural selection, a key component of evolutionary theory, involves three main ingredients:

— *Genetically based inheritance of traits*—some traits are passed down from parents to offspring in genes,

— *Variation*—heritable traits vary within a population (now we know that mutation is the source of some of this genetic variation),

— *Differential survival and reproduction*—these traits will vary in how strongly they promote the survival and reproduction of their bearers.

Selection refers to the process by which environmental conditions "select" organisms with the appropriate traits to survive; these organisms will have such traits more strongly represented in the next generation. This is the basis of adaptive evolution. The insight of Wallace and Darwin was that this "natural selection" was creative - it could lead to new traits and even new species, it was based on differential survival of variable individuals, and it could explain the broad scale patterns of evolution.

Many traits that are selected for can actually hinder survival of the organism while increasing its reproductive opportunities. Consider the classic example of the peacock's tail. It is metabolically costly, cumbersome, and essentially a "predator magnet." What the peacock's tail does do is attract mates. Thus, the type of selective process that is involved here is what Darwin called "sexual selection." Sexual selection can be divided into two types:

— Intersexual selection, which refers to the traits that one sex generally prefers in the other sex, (e.g. the peacock's tail).

— Intrasexual competition, which refers to the competition among members of the same sex for mating access to the opposite sex, (e.g. two stags locking antlers).

Inclusive fitness theory, which was proposed by William D. Hamilton in 1964 as a revision to evolutionary theory, is basically a combination of natural selection, sexual selection, and kin selection. It refers to the sum of an individual's own reproductive success plus the effects the individual's actions have on the reproductive success of their genetic relatives. General evolutionary theory, in its modern form, is essentially inclusive fitness theory.

Inclusive fitness theory resolved the issue of how "altruism" evolved. The dominant, pre-Hamiltonian view was that altruism evolved via group selection: the notion that altruism evolved for the benefit of the group. The problem with this was that if one organism in a group incurred any fitness costs on itself for the benefit of others in the group, (i.e. acted "altruistically"), then that organism would reduce its own ability to survive and/or reproduce, therefore reducing its chances of passing on its altruistic traits. Furthermore, the organism that benefited from that altruistic act and only acted on behalf of its own fitness would increase its own chance of survival and/or reproduction, thus increasing its chances of passing on its "selfish" traits. Inclusive fitness resolved "the problem of altruism" by demonstrating that altruism can evolve via kin selection as expressed in Hamilton's rule:

$$\text{cost} < \text{relatedness} \times \text{benefit}$$

In other words, altruism can evolve as long as the fitness cost of the altruistic act on the part of the actor is less than the degree of genetic relatedness of the recipient times the fitness benefit to that recipient. This perspective reflects what is referred to as the gene-centered view of evolution and demonstrates that group selection is a very weak selective force. However, in recent years group selection has been making a comeback, (albeit a controversial one), as multilevel selection, which posits that evolution can act on many levels of functional organisation, (including the "group" level), and not just the "gene" level.

Middle-level evolutionary theories are theories that encompass broad domains of functioning. They are compatible with general evolutionary theory but not derived from it. Furthermore, they are applicable across species. During the early 1970s, three very important middle-level evolutionary theories were contributed by Robert Trivers.

— The theory of parent-offspring conflict rests on the fact that even though a parent and his/her offspring are 50% genetically related, they are also 50% genetically different. All things being equal, a parent would want to allocate their resources equally amongst their offspring, while each offspring may want a little more for themselves. Furthermore, an offspring may want a little more resources from the parent than the parent is willing to give. In essence, parent-offspring conflict refers to a conflict of adaptive interests between parent and offspring. However, if all things are not equal, a parent may engage in discriminative investment towards one sex or the other, depending on the parent's condition.

— The Trivers-Willard hypothesis, which proposes that parents will invest more in the sex that gives them the greatest reproductive payoff (grandchildren) with increasing or marginal investment. Females are the heavier parental investors in our species. Because of that, females have a better chance of reproducing at least once in comparison to males, but males in good condition have a better chance of producing high numbers of offspring than do females in good condition. Thus, according to the Trivers-Willard hypothesis, parents in good condition are predicted to favor investment in sons, and parents in poor condition are predicted to favor investment in daughters.

— r/K selection theory, which, in ecology, relates to the selection of traits in organisms that allow success in particular environments. r-selected species, (in unstable or unpredictable environments), produce many offspring, each of which is unlikely to survive to adulthood, while K-selected species, (in stable or predictable environments), invest more heavily in fewer offspring, each of which has a better chance of surviving to adulthood.

At a proximal level, evolutionary psychology is based on the hypothesis that, just like hearts, lungs, livers, kidneys, and immune systems, cognition has functional structure that has a genetic basis, and therefore has evolved by natural selection. Like other organs and tissues, this functional structure should be universally shared amongst a species, and should solve important problems of survival and reproduction. Evolutionary psychologists seek to understand psychological mechanisms by understanding the survival and reproductive functions they might have served over the course of evolutionary history.

While philosophers have generally considered human mind to include broad faculties, such as reason and lust, evolutionary psychologists describe EPMs as narrowly evolved to deal with specific issues, such as catching cheaters or choosing mates.

Some mechanisms, termed domain-specific, deal with recurrent adaptive problems over the course of human evolutionary history. Domain-general mechanisms, on the other hand, deal with evolutionary novelty.

EP argues that to properly understand the functions of the brain, one must understand the properties of the environment in which the brain evolved. That environment is often referred to as the environment of evolutionary adaptedness, or EEA for short.

The term environment of evolutionary adaptedness was coined by John Bowlby as part of attachment theory. It refers to the environment to which a particular evolved mechanism is adapted. More specifically, the EEA is defined as the set of historically recurring selection pressures that formed a given adaptation, as well as those aspects of the environment that were necessary for the proper development and functioning of the adaptation. In the environment in which ducks evolved, for example, attachment of ducklings to their mother had great survival value for the ducklings. Because the first moving being that a duckling was likely to see was its mother, a psychological mechanism that evolved to form an attachment to the first moving being would therefore properly function to form an attachment to the mother. In novel environments, however, the mechanism can malfunction by forming an attachment to a dog or human instead.

Humans, comprising the genus Homo, appeared between 1.5 and 2.5 million years ago, a time that roughly coincides with the start of the Pleistocene 1.8 million years ago. Because the Pleistocene ended a mere 12,000 years ago, most human adaptations either newly evolved during the Pleistocene, or were maintained by stabilising selection during the Pleistocene. Evolutionary psychology therefore proposes that the majority of human psychological mechanisms are adapted to reproductive problems frequently encountered in Pleistocene environments. In broad terms, these problems include those of growth, development, differentiation, maintenance, mating, parenting, and social relationships.

Neuroethology

Neuroethology is a synthesis of neuroscience and the study fo behaviour that explores neural mechanisms of natural behaviour. This is in contrast to other approaches to neuroscience that study the nervous system in isolation, or in the context of artificial conditions. The neuroethological approach stems from the idea that nervous systems have evolved to address problems of sensing and acting in certain environmental conditions. The functions of nervous systems, therefore, are best understood in the context of the problems they have evolved to solve.Mechanisms of sensory processing have utilised a variety of model systems, including the moth's acoustic startle response to bat sounds jamming avoidance in electric fish, or sound localisation in owls. A detailed understanding of the control and coordination of movement patters has emerged from investigations of food processing in lobsters, prey capture in frogs, or swimming in lampreys.

Sensory receptors and neurons exhibit functional tuning to a highly specific subset of stimulus conditions. For instance mechanoreceptors with sensory hairs detect air currents in crickets by transducing both direction and velocity of relevant stimulus energies. Directional tuning is determined by the preferred orientation of the hair's socket. Velocity tuning occurs when longer hairs respond to low wind velocities while shorter hairs need higher wind velocity to activate them. Each mechanoreceptor is innervated by a single sensory neuron and its directional tuning curve represents the cell's sensitivity to wind direction.

Most substrates for sensory processing show a strict spatial organsation—its topographic mapping. A system for extracting temporal information can be constructed via delay lines, a system which utilises the fact that electrical conduction in neurons is not instantaneous; it is delayed depending on the length of the axon and other properties of the neuron (e.g. myelination). The resulting delays can be used for time sensitive calculations, such as calculation of inter-aural time differences in sound localisation. A central pattern generator (CPG) is a network of neurons (or even a single neuron) which is able to exhibit rhythmic activity in the absence of sensory input.

Modern Psychology and Linguistics

Children readily imitated behaviour exhibited by other children and adults who are significant to them, especially when such behaviour is reinforced. A primary rationale for Special Education Inclusion holds that children with special needs will model their behaviour after that of nondisabled peers and thereby improve communication, motor, social and play skills.

ANIMAL BEHAVIOUR AND HUMAN SOCIETY

Many problems in human society are often related to the interaction of environment and behaviour or genetics and behaviour. The fields of socioecology and animal behaviour deal with the issue of environment behavioural interactions both at an evolutionary level and a proximate level. Increasingly social scientists are turning to animal behaviour as a framework in which to interpret human society and to understand possible causes of societal problems. Research by de Waal on chimpanzees and monkeys has illustrated the importance of cooperation and reconciliation in social groups. This work provides new perspectives by which to view and ameliorate aggressive behaviour among human beings.

The methodology applied to study animal behaviour has had a tremendous impact in psychology and the social sciences. Jean Piaget began his career with the study of snails, and he extended the use of careful behavioural observations and descriptions to his landmark studies on human cognitive development. J. B. Watson began his study of behaviour by observing gulls. Aspects of experimental design, observation techniques, attention to nonverbal communication signals were often developed in animal behaviour studies before their application to studies of human behaviour. The behavioural study of humans would be much diminished today without the influence of animal research.

Charles Darwin's work on emotional expression in animals has had an important influence on many psychologists, such as Paul Ekman, who study human emotional behaviour.

Harry Harlow's work on social development in rhesus monkeys has been of major importance to theories of child development and to psychiatry. The work of Overmier, Maier and Seligman on learned helplessness has had a similar effect on child development and psychiatry.

The comparative study of behaviour over a wide range of species can provide insights into influences affecting human behaviour. For example, the woolly spider monkey in Brazil displays no overt aggressive behaviour among group members. We might learn how to minimise human aggression if we understood how this species of monkey avoids aggression. If we want to have human fathers be more involved in infant care, we can study the conditions under which paternal care has appeared in other species like the California mouse or in marmosets and tamarins. Studies of various models of the ontogeny of communication in birds and mammals have had direct influence on the development of theories and the research directions in the study of child language. The richness of developmental processes in behaviour, including multiple sources and the consequences of experience are significant in understanding processes of human development.

Understanding the differences in adaptability between species that can live in a variety of habitats versus those that are restricted to limited habitats can lead to an understanding of how we might improve human adaptability as our environments change.

Research by animal behaviourists on animal sensory systems has led to practical applications for extending human sensory systems. Griffin's demonstrations on how bats use sonar to locate objects has led directly to the use of sonar techniques in a wide array of applications from the military to fetal diagnostics. Studies of chimpanzees using language analogues have led to new technology (computer keyboards using arbitrary symbols) that have been applied successfully to teaching language to disadvantaged human populations. Basic research on circadian and other endogenous rhythms in animals has led to research relevant to human factors and productivity in areas such as coping with jet-lag or changing from one shift to another.Research on animals has developed many of the important concepts relating to coping with stress, for example studies of the importance of prediction and control on coping behaviour.

ANIMAL BEHAVIOUR AND THE ENVIRONMENT

The behaviour of animals often provides the first clues or early warning signs of environmental degradation. Changes in sexual and other behaviour occur much sooner and at lower levels of environmental disruption than changes in reproductive outcomes and population size. If we wait to see if numbers of animal populations are declining, it may be too late to take measures to save the environment. Studies of natural behaviour in the field are vital to provide baseline data for future environmental monitoring. For example, the Environmental Protection Agency uses disruptions in swimming behaviour of minnows as an index of possible pesticide pollution.

Basic research on how salmon migrate back to their home streams started more than 40 years ago by Arthur Hasler has taught us much about the mechanisms of migration. This information has also been valuable in preserving the salmon industry in the Pacific Northwest and applications of Hasler's results has led to the development of a salmon fishing industry in the Great Lakes. Basic animal behaviour research can have important economic implications. Animal behaviourists have described variables involved in insect reproduction and host plant location leading to the development of non-toxic pheromones for insect pest control that avoid the need for toxic pesticides. Understanding of predator prey relationships can lead to the introduction of natural predators on prey species. Knowledge of honeybee foraging behaviour can be applied to mechanisms of pollination which in turn is important for plant breeding and propagation. An understanding of foraging behaviour in animals can lead to an understanding of forest regeneration. Many animals serve as seed dispersers and are thus essential for the propagation of tree species and essential for habitat preservation.

The conservation of endangered species requires that we know enough about natural behaviour (migratory patterns, home range size, interactions with other groups, foraging demands, reproductive behaviour, communication, etc) in order to develop effective reserves and effective protection measures. Relocation or reintroduction of animals (such as the golden lion tamarin) is not possible without detailed knowledge of a species' natural history. With the increasing importance of environmental programmes and human management of populations of rare species, both in captivity and in the natural habitat, animal behaviour research becomes increasingly important. Many of the world's leading conservationists have a background in animal behaviour or behavioural ecology. Basic behavioural studies on reproductive behaviour have led to improved captive breeding methods for whooping cranes, golden lion tamarins, cotton-top tamarins, and many other endangered species. Captive breeders who were ignorant of the species' natural reproductive behaviour were generally unsuccessful.

ANIMAL BEHAVIOUR AND ANIMAL WELFARE

Our society has placed increased emphasis on the welfare of research and exhibit animals. Law now requires attending to exercise requirements for dogs and the psychological well-being of nonhuman primates. Animal welfare without knowledge is impossible. Animal behaviour researchers look at the behaviour and well-being of animals in lab and field. Many in our society are concerned with scientific illiteracy, the lack of interest that students have in science and the fact that women and minority groups are underrepresented in science. Courses in animal behaviour and behavioural ecology serve as hooks to interest students in behavioural biology. At the University of Wisconsin, Madison more than 700 students a year take courses in animal behaviour and behavioural ecology in the Departments of Anthropology, Psychology and Zoology, yet none of these courses serve as required courses for majors. Cornell University enrolls nearly 400 students in an Introduction to Behaviour course that is required of only 60-70 students. Enrollment has grown by 30% in the last three years. At the University of Stirling, Scotland, 75% of graduates in Psychology enroll in the elective, non-required animal behaviour course. At the University of Washington, Seattle, more than 300 students enroll each quarter in a basic animal behaviour class. Similar results can be found on many other campuses.

BEHAVIOUR AND THE DETECTION OF ANIMAL DISEASE

Observation of animal behaviour provides a great deal of information about the extent to which the animal copes with the environment in which they are kept. It can also contribute to the design of better housing and management systems for farm animals. An obvious example is climate. Temperatures which are either too hot or too cold will lead to behavioural changes which are readily detectable. If dairy cows are standing in free stalls rather than feeding or lying down it probably indicates that the stalls are uncomfortable. If the environment of pigs is very barren and unstimulating this will lead to some aberrant behaviours such as aggresion and also to stereotypes. Domestic animals display the same ranges of behaviours that would be seen in other animals and have both instinctive and learned behaviours. Understanding domestic animal behaviour is not only fascinating and intellectually stimulating it is also crucial in their management, productivity and welfare. Often the experienced veterinarian or owner who has a sound understanding of normal animal behaviour can identify the existence of a problem from a change in posture, sounds etc. Some are very easy to recognise such as acute laminitis others can be quite subtle. An animal kicking at its belly probably has an abdominal pain, while sweating in horses is a sign of acute pain e.g. colic. Dogs with ear infections often tilt their heads. Excessive grooming or scratching is indicative of parasites or irritations. A change in the order in which an animal comes into the milking parlor, particularly if a dominant cows hangs back, may indicate illness.

Control of Breeding and Feeding

The definition of a domestic animal is one which has been altered by selective breeding and control of the food supply by humans. An understanding of behaviour allows the detection of animals in heat, can be used to identify feeding problems, modify maternal behaviour to accept orphaned animals, manipulate group numbers and size of animals to reduce aggresion and so on.

Training

In training domestic animals heavy reliance is placed upon associative conditioning i.e., positive reinforcement of desired behaviours, but the genetic blueprint is also very important. For example the genetic predisposition of the border collie to herd is adapted by consistent training to increase skills and control of that instinctive ability. This is a particularly crucial aspect of animal science with respect to the companion and recreation species. The use learning theory is also applied to help animals relearn to reduce vices and bad habits.

REFERENCES

Alcock, J. 2001. *Animal Behaviour: An Evolutionary Approach* (7 ed.) Sinauer Associates Inc., Massachusetts. pp. 341.

Bernstein DA et al. 2003. *Psychology*, 6th ed., Houghton Mifflin Co.: Boston.

Grier JW & T Burk, 1992. *Biology of Animal Behaviour*. Mosby Year Book, St. Louis.

Sherrington CS. 1899. *On the Spinal Animal* (The Marshall Hall Lecture). Medico-Chirurgical Transactions, London 82: 449-477.

Wilson EO. 1979, *On Human Nature,* Harvard University Press.

2

Evolution and Animal Behaviour

Students of animal behaviour have been applying concepts from ethology to human behaviour for decades. Tinbergen and Lorenz saw ways in which their concepts applied to humans, and their ideas were popularised by authors like Desmond Morris, who wrote a bestseller in 1967 called The Naked Ape, full of speculations about how various species-typical behaviours of humans might have evolved.

The 1990s saw the publication of many books and articles using the phrase evolutionary psychology. Like sociobiology and human ethology, this discipline interprets human behaviour as a manifestation of underlying neural circuitry modified by evolution. Evolutionary psychologists explain common tendencies of humans by analysing their adaptive value. Topics studied by evolutionary psychologists include the following:

1. *Factors influencing mate selection.* Researchers have found that people generally find statistically average faces more attractive than faces with extreme features. Humans (and animals) prefer symmetry of the body and face in a potential mate. Not surprisingly, people also like healthy skin and consider it more attractive than blemished or diseased skin.
2. *Nonverbal and largely automatic forms of social communication.* Humans, like other animals, have species-typical displays. Eibl-Eibesfeldt studied the eyebrow flash that serves as a recognition signal and invitation to interaction. Eibl-Eibesfelt verified an observation made by Margaret Mead, the anthropologist: that young girls of many cultures use a stereotyped flirtation display.
3. *Simulations and models of evolutionary processes.* Some psychologists use mathematical or computer models to predict how individuals with particular traits could interact in large populations. The impact of traits on the population as a whole can be modeled on a computer, given certain starting assumptions. Game theory can be used to study social cooperation or competition, negotation or predation.

4. *Adaptive function of things we take for granted.* What is the adaptive function of a yawn? For primates, it is a dominance display. However, it also cools the brain. Competing explanations sometimes lead to competing predictions that can be tested with research, or it could be (as in this example) that a behaviour turns out to have multiple functions.

5. *Evolutionary explanations of strange or repugnant events.* Honor killings and jealous rages by spurned males may have "worked" in the ancestral environment. Women who stay with an abuser may be acting out an ancient script that was once genetically advantageous. The Stockholm Syndrome (in which kidnapped people bond with their captors) may have been an adaptive response in ancient times.

Speculations about evolution and behaviour are, in a sense, too easy. Anybody can conjure up explanations for why a currently-existing pattern is adaptive. It is all *post hoc* or after-the-fact (the behaviour exists, therefore it must be adaptive), and nobody can disprove it. Are there better ways to test the claims of evolutionary psychologists?

Authors from the early to mid 20th Century, such as Carl Jung, Tinbergen, Lorenz, Eibl-Eibesfelt and popularisers like Morris, relied on universality as their main argument. If a pattern (such as the eyebrow flash, or attraction to made-up faces) appeared in culture after culture, in widely separated places on earth, then it might come from the human biological blueprint. That is still a compelling argument in the case of truly ubiquitous (universal) human behavioural patterns. Presumably they are ubiquitous for a reason.

Now that DNA can be mapped and altered on a location-by-location basis, behavioural genetics (the field that relates behaviour to genes) is becoming more precise and definitive. If knocking out or introducing a gene produces a reliable consequence in behaviour, then this is clearly a point at which evolutionary change could be made through mutation or recombination.

Another avenue of change is through environmental influences that modify gene *expression* while leaving the underlying DNA untouched, an effect called *epigenetics*. Epigenetic factors include dietary influences on obesity, activity level, and health. Epigenetic factors may be passed on to children, so this is a whole new level of complexity to consider for modern scientists studying genetic influences.

In his classic study of the Yanomami Indians of the Amazon, Chagnon reported that warriors (despite the risk of injury and death) reproduced at a higher rate than men who chose not to fight. Warriors were preferred as mates, married more often, and produced more children. This finding is controversial, but it is an example of an evolutionary argument relevant to behaviour and backed by replicable evidence that could be studied directly or in computer simulations.

Classic findings about non-verbal social communication lend themselves to laboratory study by psychologists, often in college settings using subject pools drawn from modern student populations. For example, students can be asked to rate faces for attractiveness. This type of research can be used to document widespread human preferences for such things as facial symmetry, "average" features, large pupil size, and clear healthy skin, although researchers must always take care not to overgeneralise from student populations.

The effects of different strategies of competition or cooperation can be modeled on computers or simulated in laboratory games pitting one participant against another. The effect of genetic costs or benefits on the population numbers of a species can be simulated on a computer and the effects of varying assumptions about the ecosystem can be tested.

With all the solid research supporting evolutionary influences on behaviour, it is perhaps no wonder that evolutionary perspectives are achieving widespread acceptance within the discipline of psychology. However, students of evolutionary psychology should keep in mind several cautionary points:

What are a few complications in relating genes to behaviour?

— Almost all behaviour is the result of *multiple* genetic influences (i.e. most behaviour is *polygenic*).

— A single genetic change almost always results in *multiple* behavioural changes. This is called *pleiotropy*. It is important because *all* the changes produced by a genetic alteration (not just the most obvious) can affect evolution.

— A behaviour that evolved in the past to serve one function may, in later times, serve entirely different functions.

The last point is sometimes called *evolutionary opportunism*. Evolution always works with *what is already there*. A behaviour or body part that is already existing may have previously unexploited uses or benefits. If so, it can fall under a new set of selective pressures.

For example, ground-dwelling birds may use a *pre-flight* movement in their mating displays, even though members of the species can no longer fly. Some time in their ancestral past, the pre-existing behavioural pattern (pre-flight movement of the wings) was assimilated into the mating display. Why? Because it was available in the male bird's behavioural repertoire, and some females liked it. Now it serves a new function: attracting a mate.

Consequently, the genes that control muscle development and expression of that behaviour are maintained by new selective pressures. In other words, the behaviour serves a new purpose (helping to attract a mate) and that is what keeps a flight movement programmed into the species DNA of a flightless bird. Such a behaviour is said to be *emancipated* from its original function.

The twin concepts of *opportunism* and *emancipation* greatly complicate the interpretation of evolved behaviour. One cannot assume the present function of a behaviour (or body alteration) is the same as its ancestral function.

For example, *dreaming* may have evolved to serve some basic biological purpose unrelated to cognition, but in humans the vivid nighttime experiences may have influenced daytime behaviour. If the influences were positive (aiding differential reproduction) then dreaming may have taken on a new function for humans, such as helping with problem solving. In that case, the occurrence of meaningful dreaming could be an example of opportunism: emergence of a new function for an existing behaviour.

Dreaming is probably not *emancipated* from its original function, because it probably still serves its original purpose as well as occasionally offering guidance for decision-making. But the newer function of solving problems (and giving warnings, etc.) may have emerged well after the original function of dreaming in pre-human mammals. Warnings and inspirations from dreams can have considerable impact on reproductive success, so dreaming may now be adaptive for humans in several different ways: aiding memory, guding decisions, keeping people out of trouble during night hours, and many other possible functions.

The above factors are all complications of evolutionary thinking. Keeping them in mind can help a student avoid simplistic thinking. In addition, students should avoid several illogical ways of discussing evolution.

Evolution should not be reified or treated as a thing or a force of nature. Evolution is just a word used to label all the various factors that influence differential reproduction and propagation of DNA. Evolution is not a thing or an agent with a separate existence. Therefore it is erroneous to speak of evolution *making something happen*. That type of language is a sloppy shorthand for saying that a genetic change leads some creatures to prosper and reproduce while others fail to pass on their genes.

Evolutionary change may be "blind" (unpredictable) but it is not *random* in a statistical sense. Consider how babies resemble their biological parents. The outcome is not predictable, but neither is it random.

In statistics, a random process in one in which all outcomes are equally likely. However, when a biological system is altered (even by a random process like a cosmic ray) the results are not random in a statistical sense. The evolutionary past of any system makes some outcomes much more likely than others. The range of possible outcomes (when genetic expression is altered) always depends on which components and arrangements *already exist in the system*. Those pre-existing structures have been cumulating and modifying for millions of years. (In the same sense, a human never has a truly "random" thought...all thoughts are influenced by a person's history of learning, etc.)

At the same time as the details of an evolutionary process are unpredictable, the *general shape or outcome* of an evolutionary process *can* be predicted, and when a given mutation or behaviour is sufficiently adaptive, it may emerge independently several different times. Therefore one cannot assume that similar structures or behaviours are produced by the same ancestral process. Sometimes identical behaviours in different population grooups are *analogous* (resulting from similar evolutionary processes) rather than *homologous* (resulting from a common ancestors). One cannot distinguish analogous vs. homologous processes on the basis of observation alone. DNA analyses are required to resolve the issue of common ancestry.

Convergent evolution is the term used to label strongly constrained, repeated evolutionary outcomes. Some adaptations appear more than once, as the result of independent processes that differ in their details. An adaptation will appear very reliably if there is a sufficiently compelling network of constraints or selective pressures.

When humans starting domesticating cattle, for example, it became highly adaptive for adult humans to tolerate lactose, so adults could benefit from drinking milk from their cattle. This was a life or death matter during famines and droughts. Lactose tolerance in adulthood required only a simple genetic modification to preserve an enzyme already produced in babyhood. Consequently, as shown by DNA evidence, lactose tolerance evolved five different times in the past 2-3,000 years, independently, in human populations of northern Africa. That is an example of convergent evolution.

Ironically, the *parallel between learning and evolution* has been discovered repeatedly by different scientists. Pringle, Ashby, Donald T. Campbell, Jean Piaget, and B.F. Skinner all noticed it. Both learning and evolution proceed by variation and selective retention of adaptive variants, the evolutionary pattern. It is certainly fitting that this insight occurred independently to so many scientists, because it emphasises the parallel between learning and evolution itself. Convergent evolution can occur in scientific discoveries (a form of learning) just as it can in biological evolution.

EVOLUTIONARY BIOLOGY

Scientists have long sought to view the great diversity of organisms as a collection of distinct units. The species, as the atomic unit of diversity, represents a group of interbreeding natural populations that are reproductively isolated from other such groups. When individuals breed offspring, the genes of individuals are shuffled within a common gene pool representing the species' identity. The identity of species is based on the ability to breed, rather than on physical similarity. Limited transfer of genes between species causes different species to take on specific appearances and characteristcis.

Selection and Microevolution

Individuals will generally show great differences in the number of offspring they mange to contribute to the populations next generation. If differences in individual genotypes affect the bearer's reproductive success, then the frequencies of the genotypes will change over generations; genotypes with higher fitness become more common.

Fitness

Fitness describes an individual's ability to reproduce successfully. relative to that of other members of its population. It measures changes in the proportion of genes from one generation to the next. Genes associated with higher fitness become more common while reduced fitness will decrease theit proportion. Ftness of a genotype is an averaged quantity which reflects the reproductive outcomes of all individuals with that genotype. When fitness measures the relative quantity of gene copies present in individuals of the next generation, they may have arrived there in one of several different ways. An individual may show greater reproductive success becayse it itself reproduced, or it may have helped relatives with similar genes to reproduce.

Artificial Selection and Breeding

Consider the work of a professional dog breeder. Lets consider a person trying to obtain a breed that is best suited for hunting of water fowl. The breeder specifically chooses individuals which most closely match the desired traits and selectively breeds them to each other in the hope of obtaining progeny with a combination of useful traits. Such plans can only succeed if:

— phenotypic heritability in morphological and behavioural features is coded by <genes> (e.g. webbed toes, willingness to swim at least partially controlled by genes)
— phenotypic variability is present (some have a particular characteristic, some don't)
— differential survival/reproductive success of phenotypic variants is produced (you pick the ones who do)

Repeated, selective breeding events will alter the proportion of different genes over time. Genes which were present in those individuals that the breeder selected for reproduction will become overrepresented while those that occurred predominantly in discarded individuals will decrease in proportion. Animal breeding is a slow process, however, a combination of strong selection and a high degree of heritability can change the relative abundance of genes in a population up to 10% per generation.

Not all of the selections that we make are done intentionally. Penicillin became widely available during the second world war for the treatment of infected wounds. Just four years after drug companies began mass-producing penicillin in 1943, microbes began appearing that could resist it. With frequent and indiscriminant use of antibiotics we have fostered the emergence of antibiotic resistance in a variety of microorganism.

The ability to withstand the effects of an antibiotic occurs when a rare mutation renders a small subset of individuals with lowered sensitivity to the effects of the drug. During the course of antibiotic treatment the wildtype individuals are killed first as intended while the mutants are able to survive a bit longer. If the treatment is stopped before the drug had an opportunity to kill all pathogens regardless of slight differences in sensitivity to it, only resistant individuals will survive and be able to infect new hosts. With each antibiotic treatment that ends prematurely we unintentionally select for those individuals that exhibit a higher ability to tolerate the drug.

Natural Selection

Natural selection is the process by which favorable traits that are heritable become more common in successive generations of a population of reproducing organisms, while the proportion of less successful traits shrinks. "Favorable" and "successful" are defined in a purely functional sense as environmental conditions allow some individuals to leave more offspring than others. Selection cannot act on the genes *per se* but rather works on their expression into observable characteristics. Individuals with favorable phenotypes are more likely to survive and reproduce then the genotypes associated with favorable phenotyps will increase in frequency with each generation. Over time, this process can result in the emergence of adaptations that specialise a group of organisms for a particular ecological set of conditions (i.e., microevolution). If different subsets of the population adapt to excell in one of multiple, distinct niches, the emergence of any reproductive barriers between the groups may allow the population's split into separate species (i.e., macroevolution).

More offspring are produced than finite resources can support. Individuals thus can be viewed as in a constant struggle for existence. Individuals within a population are rarely clones, they commonly show variation in phenotypes as well as genotypes. Some of this variation in behavioural phenotypes is heritable. As successful variants are more likely to survive and reproduce, their genotypes will be become overrepresented in the next generation.

Density-independent and density-dependent growth models: Exponential Model: a species can potentially increase in numbers according to a geometric series — Thomas Robert Malthus (1766-1834) Logistic Model: the rate of population increase may be limited, i.e. it may depend on population density — Pierre Verhulst (1838). Carrying Capacity (K): an environment's maximum persistently supportable load.

Natural selection is the process by which environmental effects lead to varying degrees of reproductive success among individuals of a population of organisms with different hereditary characters, or traits. The characters that inhibit reproductive success decrease in frequency from generation to generation. It is the process whereby certain genes (alleles) gain greater representation in the following generations compared to other alleles. Adaptations are the complement of traits that increases the fitness of the owner. An individual's Fitness or Reproductive Success is the relative probability that an animal of a particular genotype and phenotype will manage to contribute its genes to the next generation.

A major concern of animals and other critters is to protect themselves from predators in order to survive and reproduce and pass their genes off to a new generation. Many animals have evolved adaptations known as antipredator devices such as camouflage and chemical toxins.

Animals use camouflage to blend in with their environments in an attempt to be unrecognisable by predators. Other organisms such as the monarch butterfly contain chemical toxins that are secreted into the predator's mouth when it attempts to eat the butterfly. The monarch butterfly also has warning colouration that gives a warning sign to predators to remind them that the butterfly is toxic and should not be eaten.

These antipredator devices are so successful that other organisms have been known to mimic them. The organism that is mimicked is known as the model and the third party that is deceived by the model and its mimic is known as the receiver. The mimics have learned to take advantage of the colour patterns and markings that predators have learned by experience to avoid. The model is usually a species that has an abundant population and has successfully warded off predators with an antipredator device.

Organisms have learned to mimic their surroundings or environment in an attempt to "hide" from predators. For example, lizards have learned to mimic tree trunk colour which proves to be very successful as predators will simply move past them as they believe that they are simply looking at a tree. Another example of this type of mimicry can be seen with the Katydid who will mimic a leaf in both colour and shape in an attempt to be hidden.

Some prey animals have evolved certain patterns on their bodies that mimic other animals in an attempt to startle their predators. The most common example of this type of mimicry can be found in some moths and butterflies who flash eye spots on their wings to predators. These eye spots startle the predator who believes that the eyes belong to a much larger animal that may be a threat to them.

In one form of mimicry known as aggressive mimicry, an organism will mimic a signal that is either deceptive or attractive to its prey. One example of this involves the praying mantis who will mimic flowers to attract insects that they can then capture and eat. Organisms can also imitate the behaviours of other organisms. Moth caterpillars, for example, will imitate the motion and body movements of a snake in order to scare off predators that are usually a prey item for snakes.

One of the most popular types of mimicry involves the warning colouration found on inedible or toxic organisms such as the monarch butterfly. Once these toxic organisms have adapted this warning colouration which warns predators to stay away, other organisms may start to mimic this warning colouration in an attempt to stay alive.

Batesian mimics are those mimics that imitate unpalatable species even though they are palatable. Therefore, one species is harmful while the other is harmless. The wasp is a great example of Batesian mimicry. The wasp is the model species in this example as it possesses a sting which enables it to escape from predators. The bright warning colouration of the wasp has been mimicked by many other insects. Even though the mimics are harmless, the predator will avoid them due to bad experiences with wasps with the same colouration. With Mullerian mimicry, many unpalatable species share a similar colour pattern.

Mullerian mimicry proves to be successful as the predator only has to be exposed to one of the species in order to learn to stay away from all the other species with the same warning colour patterns. The black and yellow striped bodies of social wasps, solitary digger wasps, and caterpillars of the cinnabar moths warn predators that the organism is inedible. This is a great example of Mullerian mimicry as all of these unpalatable, unrelated species have a shared colour pattern that keeps predators away.

Mimicry is a very successful antipredator device that species have evolved over many generations. As one can see, organisms have come to mimic many different characteristics such as colour patterns and behaviours. However, selection only favors the mimics when they are less common than the model. Therefore, the fitness of mimics is "negatively frequency-dependent."

Industrial Melanism of Peppered Moths

Within one hundred years (1850 to 1950); the dotted whitish form of the peppered moth (*Biston betularia*) was almost entirely replaced by the melanic (black) form. The melanic form appeared to be best suited for survival against the soot that had collected on forest tree trunks from pollution because the moths could camouflage with their resting area on the tree. The dotted whitish form of the peppered moth was no longer predominant in this environment because they were easily detected and predated on. How does this happen?

Many animals have anti-predator adaptations. Adaptations are defined as "a heritable trait that either spread because of natural selection and has been maintained by selection to the present or is currently spreading relative to alternative traits because of natural selection"

Anti-predator adaptations suggest that a heritable trait, which enables the organism to hide from predators by seeking cover against a background, has spread by natural selection because of reproductive success. H.B. D. Kettlewell's experiments on the peppered moths, as well as, those conducted by R.J. Howlett and M.E.N. Majerus have proven that the peppered moth's preference for their resting places on trees are anti-predator adaptations.

In 1955, H.B. D. Kettlewell published his study on pepper moths: *Selection Experiments on Industrial Melanism in the Lepidoptera*. Kettlewell hypothesised that the dotted whitish form of the peppered moth were more likely to be eaten than the melanic form because they could be easily detected against the soot covered trees. His studies showed that the moths that were easily identified by humans were at a higher risk of predation from birds. The dotted whitish form was at higher risk of predation than the melanic form in the polluted environment.

Figure 1. Peppered Moth: Dark and Dotted Whitish

Howlett and Majerus further examined this hypothesis in their study: *The Understanding of Industrial Melanism in the Peppered Moth*, published in 1987. They tested it by pinning 50 of both forms of the peppered moths on pale and dark tree trunks.

Calculations:

Moths in Polluted (dark) Woodland

Dotted Whitish Form: (30/50) x 100 = 60% Melanic Form: (20/50) x 100 = 40%

60%—40% = 20% The Dotted Whitish form is predated on 20% more than the melanic form in the polluted woodland.

Moths in Non-polluted (pale) Woodland

Dotted Whitish Form: (15/50) x 100 = 30% Melanic Form: approx. (30/50) x 100 = 60%

60%—30% = 30% The Melanic form is predated on 30% more than the dotted whitish form in the nonpolluted woodland.

Their studies showed that the dotted whittish form was preyed on 20% more than the melanic form in the polluted woodland. The melanic form is predated on 30% more than the dotted whitish form in the nonpolluted woodland.In addition to the difference in predation, their studies also showed that in both the polluted and nonpolluted environments the moths that were located on the limb joints instead of the trunks were less likely to be preyed on.

Howlett and Majerus showed that the dotted whitish form was at a higher risk of predation when resting against the polluted (dark) trees. The melanic form had a greater chance of survival and reproduction because they were less likely to be detected against the soot-covered trees, the same findings of Kettlewell years before. Their experiment proved that the melanic form was an anti-predator adaptation, which is why the dotted whitish form had become so rare. The melanic form had become the dominant trait for survival and reproduction in the polluted woodland.

Antipredator Adaptations used by the Monarch Butterfly

The relationship between prey and predators continually changes. Prey need to find ways to outsmart predators in order to survive and reproduce. One way monarch butterflies increase their fitness is by forming huge groups of up to ten million. Increasing group size lowers the probability that any one monarch butterfly will be captured. This is known as the "dilution effect." Moreover, butterflies located in the center of a large group are more likely to survive than those on the outside.

Monarch butterflies that choose to migrate to closed-area overwintering sites are less likely to be attacked by a predator. Also, by reacting as a group to the movement of a predator, monarch butterflies are better able to scare away predators. This "mass startle effect" is thought to stun the predators and provide time for the butterfly to escape (true shock and awe indeed). The aforementioned are collective ways in which the butterfly behaves in order to elude a predator. There are, however, certain individual inherent features that the monarch butterflies possess that increase their probability of avoiding a predator.

Figure 1. Monarch Butterfly

Monarch butterflies contain chemicals that are toxic to many predators. Evidently, this makes other similar but harmless species envious. In Batesian mimicry, a palatable species attempts to mimic an unpalatable species in an attempt to increase its own fitness. The monarch butterfly species is one that some Batesian mimics model themselves after perhaps because the monarch butterflies are so successful at avoiding predators.

The Batesian mimics, although they are not harmful to predators, experience increased fitness because they model a potentially harmful species such as the monarch butterfly. The viceroy butterfly (*Limenitis archippus*) is one such butterfly that models itself after the monarch. Although monarch butterflies do not gain an increase in fitness as models of Batesian mimics, they do benefit from Mullerian mimicry.

In Mullerian mimicry, two unrelated, toxic species converge on a similar morphology. If more than one unpalatable species has a similar morphological trait, then the predators may more easily recognise a Mullerian mimic as potentially harmful. This saves both the mimics and the predators time and energy.

Lastly, monarch butterflies display aposematism or warning colouration. This warning colouration is meant to be very conspicuous. Monarch butterflies make themselves conspicuous by having bright orange areas on its wings. Predators quickly learn that prey containing these bright colours are potentially harmful. For example, when a blue jay consumes a monarch butterfly, it vomits shortly after. From that point on, the blue jay associates features of the monarch butterfly, such as its bright colours, as unpalatable. Because of their morphological features, Batesian mimics, Mullerian mimics, and many other aposematic species all gain from the monarch's unpalatability.

Sexual Selection

Sexual selection, a subcategory of natural selection, was first recognised by Charles Darwin and occurs when individuals differ in their ability to compete with others for mates or to attract members of the opposite sex. By heavy courtship, fighting, or large territorial possession, males heavily compete for females. Even though a male may win a fierce competition for the mate of his choice, it is ultimately the female who decides on a partner that she wants. The female is often successful in her attempts to control reproduction by being choosy and having particular preferences for a male mate.

Females typically have a relatively larger investment in producing offspring. While a single male can often fertilise all of a female's eggs, she will not produce more offspring by mating with more than one male. By and large, a male's reproductive success increases with the number of mating opportunities he obtains, whereas a female's reproductive success is largely limited by how many offspring she can physically bear.

This results in sexual selection, in which eager males compete with each other, and females become choosy in which males to mate with. Bateman's studies on mating behaviour in fruit flies suggested that the origins of the unequal investment reside in the fact that male gametes (sperm) are cheaper to produce than eggs. Animals are therefore fundamentally polygynous, as a result of being anisogamous.

— A female can have only a limited number of offspring, whereas a male can have a virtually unlimited number, provided that he can find females willing to mate with him. Thus females generally need to be much choosier about who they mate with.

— A male can easily produce sperm in excess of what it would take to fertilise all the females that could conceivably be available [...] Hence the development of the masculine emphasis on courtship and territoriality or other forms of conflict with competing males.

— In most animals the fertility of the female is limited by egg production which causes a severe strain on their nutrition. In mammals the corresponding limiting factors are uterine nutrition and milk production, which together may be termed the capacity for rearing young. In the male, however, fertility is seldom likely to be limited by sperm production but rather by the number of inseminations or the number of females available to him... In general, then, the fertility of an individual female will be much more limited than the fertility of a male... This would explain why in unisexual organisms there is nearly always a combination of an undiscriminating eagerness in the males and a discriminating passivity in the females.

— Among polygynous species, the variance in male reproductive success is likely to be greater than the variance in female reproductive success.

— The female, with the rarest exceptions, is less eager than the male... she is coy, and may often be seen endeavouring for a long time to escape.

Intersexual Choice

Females choose mates based on many factors. One important factor is male adornments, or ornaments. For example, Marion Petrie and Tim Halliday concluded that the removal of eyespots from a peacock's tail significantly reduces his attractiveness to females. Thus, females are extremely conscious of the visual stimuli provided by males. Other factors involved in female preference in mate selection include body colouration or "gifts" the male may present to the female before copulation.

A mother is expected to always want the best for her child, even if it is a future child. Females select mates with certain traits, because they want their children to be healthy, viable, and reproductively successful. A direct and heritable benefit of mate choice by females is the assurance of bearing offspring that will survive well and display high general fitness. Non-genetic benefits of mate choice include fecundity advantages, food, parental care, or a good territory. All of the benefits, both heritable and nonheritable, ultimately lead to the greater survival of a female's offspring.

— *Passive Attraction Theory*: Sensory Bias, conspicuous signals make an individual more likely to attract the attention of a mating partner. Elaborate sensory cues alert the female that the male is reproductively superior to others. Male adornments may more readily elicit the mating responses of some females who will thus mate preferentially in favour of the adorned males.

— *Nuptual gifts*: spermatophore; territory, protection, resources; courtship indicative of parental investment. e.g., Dung beetle

— *The Good Genes Hypotheses* regards assessment of the mate's state of health as a general indicator of viability and quality. Females exhibit mate choice in order to provide their offspring with a partner's genes that will advance their offspring's chances of survival or reproductive success. The healthy mate theory occurs when females prefer males healthy enough to produce and maintain elaborate ornaments. A good example of this is in female house finches, who choose male mates based on their bright colouration. Bright colouration tells the female that the male is more resistant to pathogens and parasites. In Hamilton and Zuk's "Revealing Signal Theory" bright ornaments reveal a genuinely healthy individual in good condition. In conditions where parasites ralter male showiness and parasite resistance is largely inherited, females ought to choose those with bright colouration.

— *Zahavi's Strategic Choice Handicap Theory:* the presence of a costly trait is indicative of otherwise good underlying genes that allow an individual to prosper despite this handicap

— *Fisher's Run-away Selection*: Females that mate with attractive males are compensated for reduced fecundity by bearing attractive offspring with higher than average mating success. This "Sexy Sons Hypothesis" works by alligning the presence of a particular morphological characteristic in males with a preference for it in females.

Choosy females create a positive feedback loop favoring both males with these attributes and females that prefer them.

- *Genetic Compatibility*: MHC locus genes in human mate choice
- *The Good Parent Theory* suggests that choosy individuals ought to select partners on the basis of how well they will care for their offspring, e.g. as the female searches for a paternal male.

Inbreeding

Inbreeding refers to an elevated amount of breeding between close relatives. The resulting increase in homozygosity of the population exposes recessive, deleterious traits in homozygous form and may lead to inbreeding depression where inbred individuals exhibit reduced health and fitness and lower levels of fertility.

Outbreeding

The introduction of genetic material from unrelated individuals into a breeding line increases genetic diversity, thus reducing the incidence of disease or genetic abnormalities. It is used in breeding to restore vigor, size and fertility to a breeding line.

Sexual Selection and Ornamentation in Males

Sexual selection refers to the process by which changes in gene frequencies result from individuals that are better than others at either competing for or at attracting mates — it is the evolution of traits based on differences in mating success among individuals if:

(1) some traits increase the ability to compete with individuals of the same sex for access to mates, or

(2) some traits increase the ability to attract individuals of the opposite sex.

Ornamentation in males, which are commonly the competing sex, may result from different evolutionary forces. Sexual selection is always harsher on the competing sex.

Intrasexual Competition and interference

Dominant individuals often gain preferred access to mates, desirable territory, or other advantages that will enhance the individual's chances for transmitting its genes into future generations.

- *Precopulatory*: A common form is based on an individual's ability to physically dominate a rival. In situations where groups of mates can be readily monopolised this will likely lead to increasing size in the competing sex. Fur seal females, which rely on a small stratch of beach for giving birth, can be monopolised easily in harems.

Winning in ritualised contests, producing a louder signal, or masking an opponent's call; dominance in social groups; territorial exclusion; alternative mating strategies; sexual interference.

Postcopulatory:

- mate guarding;
- anti-aphrodisiacs;
- mating plugs; partners remain attached e.g. wolves;
- Bruce effect is a form of pregnancy disruption when a female mammal is exposed to an unknown male. Examples may also include post-implantation failure in many species of rodents.
- Infanticide
- sperm competition: e.g. primate or bat mating systems

BASIC MODELS OF MOLECULAR EVOLUTION

Adaptational Models

Adaptational Models are rooted in the view that natural selection plays the dominant role in structuring an organism's morphology, physiology, and behaviour. The extant phenotype thus is the product with adaptations that have its possessor allowed to reproduce more successfully than individuals with alternative traits.

Neutralist Models

Alternative accounts for the evolution of traits focuses on genetic drift as a primary determinant for characteristics ranging from genes to complex traits. This acknowledges a significant role for historical contingencies, where present phenotypes of population members are largely determined by the genetic makeup of a small number of founding individuals. Although this view does not deny the role of natural selection in determining the course of adaptive evolution, the theory attributes a large role to genetic drift. When one compares the genomes of existing species, the vast majority of molecular differences do not influence the fitness of the individual organism. As a result, the theory regards differentiation in such features as neither subject to, nor explicable by, natural selection. Theoretical considerations suggest that such a process is most effective in small populations and when selecting among traits that are largely adaptively neutral.

SPECIATION AND MACROEVOLUTION

Speciation is the evolutionary process by which new biological species may arise. Natural speciation has taken place over the course of evolution, although the relative importance of different mechanism in driving biodiversity remains a subject of debate. Debate also relates to the rates at which speciation events may occur over geologic time. Although some evolutionary biologists claim that speciation events have remained relatively constant over time (i.e., Phyletic Gradualism), palaeontologists such as Niles Eldredge and Stephen Jay Gould have argued that species may remain unchanged over long periods of time with speciation occurring in relatively brief intervals (i.e., Punctuated Equilibrium).

Allopatric Speciation

In allopatric speciation, two subgroups of a population become geographically isolated from each other (i.e., allopatric populations) through geographical or geological events. As these isolated populations undergo genotypic and phenotypic divergence, they are subjected to different selective pressures and may undergo genetic drift. New species form when the populations evolve independently such that they are no longer able of exchange genes when they come back into contact.

Peripatric and Parapatric Speciation

New species are formed when isolated, small peripheral populations split from the main population and occupy a novel ecological niche. Since small populations often undergo bottlenecks, genetic drift may drive the latter into significant genetic divergence (i.e., peripatric specieation). If geography only accounts for a partial separation, individuals of each species may come in contact across the barrier from time to time, but reduced fitness of heterozygotes leads to selection for reproductive barriers that prevent breeding between the two species.

Sympatric Speciation

In symptric speciation species diverge while inhabiting the same place. This may happen when a diploid cell undergoes failed meiosis, producing diploid gametes, and self-fertilises to produce a tetraploid zygote. Alternatively, subgroups of a species may become dependent on different host organisms within the same area. With reduced fitness of heterozygotes, the emergence of reproductive barriers will drive the groups towards two distinct species.

SEX AND MATING SYSTEMS

As best we can tell, life originated without sex. Although asexual reproduction would arguably be considerably less wasteful than sexual reproduction in many respects, the great majority of species have prospered with reliance on latter. in sexually repoducing organisms half of the population is lost as a reproducing unit (i.e., the cost of males). In addition, both individuals contribute only half of their genetic material to the offspring, while the other half is lost. In addition, there is the danger associated with combining your genome with a largely unknown and untrusted copy from a partner. Several possible reasons have been advanced to explain the presence of sexual reproduction:

— *Muller's Ratchet*: Mutations accumulate in asexual lineages as there is little chance to get rid of them again.

— *Vicar of Bray hypothesis*: Ability to combine individually advantageous genes allow for rapid adaptations to change

— *Red Queen Hypothesis:* By constantly reconfiguring the precise genetic set that a parasite or pathogen will face, animals with longer lifespans can provide a moving target.

Asexual lineages tend to disappear over time while sexual lineages tend to persist much longer.

Evolution of Diploidy

Most eucaryotic groups exist as diploids (i.e., those carrying 2 sets of genetic information), some occur primarily in haploid forms with a single set of genes, and others alternate between haploid and diploid phases. Different strains of yeast, which may be haploid or diploid, show advantages depending on the specific envrironmental conditions.

In a favorable setting, when nutrients are present in excess, the two strains show little difference in reproductive rates. However, when growth was limited by the concentration of a single nutrient, haploid strains of yeast were able to out-compete diploids. This is consistent with the idea that haploid cells are more efficient as simple replicators.

The main advantage of diploid cells emerges with a greatly improved resistance to damage. Sexual haploids may combine the advantages of both: spending much of the life cycle in the haploid state, then temporarily fusing to become diploid, followed by splitting to the haploid state.

Diploid organisms appreciate several key advantages:

— A second set of genetic information provides a backup in case of somatic mutations. With two copies of every gene, a diploid organism is able to mask recessive deleterious mutations. DNA damage can be repaired in a diploid state, robinsince there are two copies of the gene in the cell and one copy is presumed to be undamaged.

— A second set provides one copy as a substrate that is free to be altered by mutation, while a second, trusted copy remains available for basic cell functionality
— Increases genetic diversity that may or may not be expressed
— Provides conditions that allow for sexual reproduction.

Evolution of Anisogamy

Disruptive selection in gamete characteristics underlies the evolution of Anisogamy. Anisogamy is the condition in which one type of gamete becomes increasingly large (i.e., the egg) to provide the zygote with storage for a head start in development. As size increases, gametes becomes less mobile. In turn selection produces another type of gamete (i.e., sperm) that is optimised for locomotion to find and fuse with the egg. The larger gamete is termed the female sex, the smaller gamete is that of the male. Compared to isogamy this allows for both increased provisions for zygote and optimised mobility.

Bateman's Principle: increased reproductive investment is accompanied by increased selectiveness for mating partners. While female gametes are energetically expensive, large and limited in number, male gametes are cheap, mobile and available in large numbers. This asymmetry leads to reluctant females as the choosy sex, while ardent males are the competing sex.

A.J. Bateman phrased this rule based on work where he had examined variation in reproductive succcess between male and female fruitflies. He had placed 3 male and 3 female fruitflies into a container and subsequently assessed the reproductive success of all individuals. Most females in contrast had reproduced with little variation, while only a few males had accounted for all reproduction and most males had no success at all.

Limits to reproductive success thus differ between sexes where males are generally limited by the number of successful matings, while that of females is primarily determined by the rate of egg production. This favors behavioural traits for choosy females and ardent males. Differential reproductive investment will polarise operational sex ratio (i.e., most females and only a few males get mating opportunities).

The Coolidge effect describes phenomena whereby males show high sexual performance given the introduction of previously unknown, receptive females.

Trivers-Willard Hypothesis: Moreover, as only the most competitive males achieve mating success, a bias in sex ratio may emerge where dominant/healthy females produce a relative excess of males which stand a disproportionate chance to develop into the successful male. Female offspring are the safe bet, while males may either be highly suceessful or a spectacular bust.

Evolution of Sex Ratio

Sex ratio describes the relative number of males to females in a population.

The primary sex ratio refers to the number of males vs females at conception. The distribution is most commonly around 50:50. Such a sex ratio might not seem very efficient considering that a male can fertilise several females. Fisher's theoretical work illustrated that such a balance forms a general equilibrium point. If there are fewer males than females, then males face better odds in mating. An advantage for males favors females who produces extra sons. The same argument follows for deviations in the opposite direction. Thus, if a population ever comes to deviate from a ratio of 50:50, natural selection will tend to drive it back to that balance.

The secondary sex ratio, the ratio at time of birth, in humans usually shows a slightly bias towards boys (p=0.53%). With a somewhat higher mortality of males the tertiary sex ratio, the ratio of mature organisms, becomes increasingly skewed towards females.

The operational sex ratio refers to the relative number of receptive males per receptive female in the population. The operational sex ratio is usually biased heavily towards males indicating that females have many more males to choose from than the other way round.

Mating Systems

Several types of Mating systems can be recognised:

- Promiscuity
- Monogamy (one male, one female partner)
- Polygyny (one male, multiple female partners)
- Polyandry (one female, multiple male partners)
- Polygynandry (multiple female, multiple male partners; Promiscuity)

Mating systems are highly fluid evolutionary entities. They are dynamic and highly optimised systems where many different factors figure into the equation. Selection will always foster the most successful tradeoffs for these parameters, but many strategies may lead to success. Mating partners are rarely selected at random. In most species it is strongly controlled by diverse factors. Although many individual strategies may prove successful, a variety of considerations provide predictive power for the type of mating system that is present:

Reproductive Strategies

Reproduction is the sole method through which a species may continue, and reproduction manifests itself in numerous deviations throughout nature. The ultimate goal of each species is to produce the maximum amount of offspring while exerting the least amount of energy. Predation pressures, resource accessibility, and competition with the species for attracting mates greatly influence the likelihood of an individual within a species producing viable offspring and donating genes to subsequent generations. Different fish species may be observed in order to learn the unique reproductive strategies employed throughout nature from fertilisation tactics to parental care methods that have adapted these fish in continuing their species.

Females typically exert much more energy and time into reproduction than males, and therefore, females are usually more selective in choosing mates. Females typically favor males that are larger in size, possess more elaborate physical traits than the other males, and display more energy in courtship activity. However, smaller and less desirable males must still be able to fertilise the eggs, or, according to natural selection, they would have been eliminated from the population.

Although slightly more unsettling, sexual coercion results in the fertilising of eggs just as sexual courtship does. The term "sneakers" refers to smaller males who opt to race into the nest more desired males and fertilise the eggs while the nesting male pursues the female. The term "satellites" describes cryptic males mimicking females in appearance and behaviour so that they are able to fertilise the eggs while the unsuspecting couple spawn.

Although fish species exhibit both internal and external fertilisation, fish also exemplify many unique strategies for parental care. Most fish species utilise broadcast spawning, which is a method a releasing gametes into the water and providing no parental care. The goal of this reproductive method is to produce the maximum amount of progeny in hopes that as many offspring as possible will survive. Many fish species, however, opt to employ parental care in order to supply their smaller number of offspring a greater chance of survival. Oral brooding and sex role reversal are two unique methods of parental care that have proved successful in certain fish species.

Oral brooding, although quite rare in nature, is a parenting method adopted by many fish in the Family Cichlidae. After the successful fertilisation of the eggs, Cichlids, such as Tilapia, place the egg clutch into the mouth of the female. The fecundity is significantly lower in oral brooders; however, the eggs tend to be larger and receive more nutrients. By placing the eggs into the oral cavity, the female is able to provide protection for the eggs, and the churning ability of the female rotates the eggs so that each egg is exposed to oxygenated water.

Similar to oral brooding, sex role reversal in fish is rare but is utilised in the Family Sygnathildae. In sea horses, the female inseminates the male by inserting the oviduct into the male brooding pouch several times to ensure fertilisation. After fertilisation is complete, the female departs, and the male attaches itself to a nearby object with its tail waiting for the eggs to mature.

Nature exhibits many alternatives that species have adapted in order to reproduce. By utilising these methods, these organisms are able to reproduce viable offspring and continue their kind. Survival of the species is the ultimate goal for all organisms, and these organisms are constantly adapting to their environment in order to do so.

Differences in Reproductive Investment

Parental investment is any behaviour toward offspring that increases the chances of the offspring's survival at the cost of the parent's. The presence of a large asymmetry in parental care paired with an ability to monopoluise mates favors conditions for male-male competition and polygynous mating systems where a small number of males account for a disproportionate amount of the breeding. As size matters as a cheat-proof arbitator, selection for increased body size and asymmetries between the sexes become common.

When resources (e.g., nesting sites, food resources) can be dominated by a small subset of males, a polygynous mating system will likely emerge:

Certainty of Paternity

Consider mammals, birds, and fish and phrase your prediction as to the sex that you would be expected to invest more into the offspring.

— external vs. internal fertilisation

— control of fertilisation site

Social monogamy is relatively rare in mammals but common in birds. Eggs, which develop internally in mammals, restrict the females ability to shift parental duties to the male. Birds, where egg development is external, is more conducive to such efforts. Differences in constraints and costs between sexes in parental care.

Polygyny

There are different types of polygyny. They are:

— *Resource-defense polygyny*: Polygyny should be more common in patchy environments with variation in territory quality. e.g., honey guides; female fitness depends on quality of resource which is controlled by the male

— *Female-defence polygyny*: e.g., seals; females aggregate at favorable sites and males monopolises access to female harems

— *Male-dominance polygyny*: e.g., crayfish
— *Scramble polygyny*: e.g., frogs, little overt competition between males
— *Lek Polygeny* is a mating system common in polygamous species of insects and birds in which the males provides no parental care to its offspring. The lek mating system is uniquely driven by the females' pursuit of their mate, rather than the males'. Males of lekking species do not hunt for receptive females. Males form aggregates in neutral locations devoid any resources valuable to females. The group of males performs intricate vocal, visual or chemical displays to lure receptive females to their lekking site. In most lekking species, these group displays typically increase the ratio of visiting females per males. At the lekking site, visiting females are able to compare the males' physiques and courtship displays, picking the most attractive male as their mate. Thus, the few, most attractive males will do the majority of the mating (about 99%), while the subordinate males do no mating at all.

Alternative Hypothesis

Kin Selection Hypothesis. The Lek Polygyny mating system promotes a heavily skewed mating success rate among lekking males. Although most of the individuals in a lek never receive a mating opportunity, lek polygamy continues to flourish among various species of birds and insects. This suggests that the fitness of subordinate males must somehow be indirectly benefited by communal displays. A number of hypotheses have been proposed to explain the reasons behind lekking behaviour. Widely recognised hypotheses include the Hotspot Hypothesis, the Hotshot Hypothesis, and the Female Preference Hypothesis. Petrie et al proposed an alternative hypothesis predicting that lekking behaviour is driven by kin selection.

In the kin selection hypothesis, Petrie et al suggests that if all the males in a lek where genetically related, then the males (even the subordinate males) would receive fitness benefits. Petrie et al demonstrated the kin selection hypothesis by determining the genetic structure of different lekking groups of the Whipsnade Park peacock population.

Peacocks, a typical lekking species, form aggregates at neutral display sites. Peacocks use their calls to attract receptive peahens. Upon the arrival of a peahen, the peacocks cease calling and perform intricate plumage displays. As common in lekking species, the peacocks with the most appealing courtship displays have a high mating success while the subordinate peacocks have no mates. The lekking sites of peacocks are carefully chosen and after a male's fourth year of lekking, he will establish his permanent lekking site. Peacocks will return to this same lekking site every mating season.

Using multilocus fingerprinting, Petrie et al compared the genetic similarity of within and between the lekking groups of Whipsnade Park's peacocks. The results supported the proposed kin selection hypothesis. As predicted, the degree of band-sharing within the leks was substantially higher than the band-sharing between the leks. The band sharing within the leks was indicative of that of half-siblings.

After this discovery, Petrie et al proposed subsequent hypotheses of why these lekking groups consisted of related individuals. Petrie el al suggested that related peacocks tended to display together because their dispersal was concentrated around their natal sites. This hypothesis was rejected when peacocks of mixed relatedness were reared away from their natal sites but, upon reintroduction into the Whipsnade Park population, joined leks with closely related peacocks. It was then suggested that peacocks' tendency to congregate with relatives was due to a shared genetic preference for habitat-selection. This was rejected due to the homogenous environmental features of the park.

Petrie et al concluded that unique structure of the peacock leks was driven by kin selection based on self-referent phenotypic matching. Petrie et al predicted that peacocks match heritable similarities of their own phenotype with those of other males. Because the authors believed that peacocks do not participate in the rearing of their offspring, they posit that a peafowl's recognition of their father must be genetically innate.

The tendency to form leks with relatives, therefore, occurs in the absence of social cues to their identity. By cooperatively forming leks with genetically related individuals, subordinate peacocks forfeit their chances of mating, but increase the chance that their genes will be passed on via successful relatives. The indirect benefits of fitness seen in peacock lekking displays outweigh the costs of communal displays.

There are four distinct genera of Peafowl: Pavo, Afropavo, Rheiinartia and Argusianus. Though it is frequentally reported that male peafowl do not participate in nest defense or rearing paternal care of young has been observed in all four genera. Only the common Indian Peafowl Pavo cristatus is frequentally kept in captivity. Its behaviours in zoo settings may not reflect its evolutionary history to the degree that wild populations of the species might.

Hotspot Hypothesis. According to the hotspot hypothesis, males form leks because females frequently visit certain "hotspots".

Hotshot Hypothesis. The hotshot hypothesis predicts that males form leks because subordinate males congregate around highly attractive males to increase their chance of being noticed by receptive females.

Female Preference Hypothesis. The female preference hypothesis predicts that males form leks because female like to visit large clusters of males consisting of a variety of potential mates from which she can quickly and safely compare the quality of her mating choices.

Monogamy

Resource considerations may foster the emergence of monogamous mating systems. When resources are difficult to dominate (e.g., a scattered renewable food source) things get more complicated.

Monogamy is a possible outcome which exists either over a lifetime (e.g., greylag geese) or serially (e.g., ducks). Also if females are widely distributed, males may not be able to monopolise them. As females may mate with another male, monogamous males may play a guarding roling towards the female. The male may derive some fitness benefits in situations where his parental contributions increased surviva; of his offspring. Under such conditions, female choice will encourage shifting part of the burden of parental care to the male.

Polyandry

Polyandry is defined as "the mating of one female with more than one male while each male mates with only one female." Exclusive polyandry (as opposed to polyandry in concert with polygyny) is very rare, occurring in only about 1% of animal populations, most being shorebirds like the sandpiper. The basis of polyandry is a sex role reversal. The females compete for the males and are larger and more colourful, while the males take on the parental role. With the sex role reversal, a natural selection against older males evolves. This is accomplished by the females tending to select the males with the best sperm in order to give the female the most offspring possible.

Younger males will more likely have fertile sperm; therefore impregnating the female on more instances than an older male with less fertile sperm. When multiple helpers are needed for successful reproduction, females specialise in egg production while pairing with multiple males in polyandry (e.g., Jacanas).

There are two major types of polyandry: simultaneous and sequential. Simultaneous polyandry is when the female controls a very large territory. Inside this territory, the female has multiple smaller nesting territories with different males. The female mates with all males simultaneously, keeping control of the smaller territories. Another form of simultaneous polyandry, cooperative simultaneous polyandry, is when the female only has one nesting area where she mates with multiple males producing a clutch of eggs of mixed parentage with all males contributing to the eggs.

Sequential polyandry, the most common form, is where the female mates and produces a clutch of eggs with one male, then leaves the male to incubate and rear the eggs while moving on to another male in a different nesting territory. Here, the female moves from one male to another, leaving the male in full responsibility of the eggs instead of sharing the responsibility.

The genetic benefits of polyandry include fertility insurance. This hypothesis suggests that by mating with multiple males, the female is guaranteed to fertilise all of her eggs. The multiple partners potentially make up for one male that may not be able to fertilise the eggs. The good genes hypothesis states that the females have multiple mates because she is in search of the male that will pass along the best genes to her offspring.

By finding this male, the female is increasing the survival rate of her offspring. The genetic compatibility hypothesis is one that suggests that the female finds multiple mates in order to find the most compatible genetic match for her eggs. While looking for a good match, she is also eliminating the males that are least compatible with her eggs.

The material benefits of polyandry can be seen through three hypotheses. The more resource hypothesis suggests that the more mates the female has, the more males she has to care for her clutch. The better protection hypothesis states that by having multiple partners, the female is better protected from predators. The infanticide reduction hypothesis is one that claims that since the female has multiple males, she has a lower infanticide rate because the males do not know which progeny belong to them. This prevents the males from killing other male's young. Although some infanticide occurs between females and other female's eggs, it is minimal among males.

Promiscuity

Alternatively, in situations where social groups are able to form, females may mate with any male member of the group leading to Promiscuity (Polygynandry). Sperm competition will in this case most likely determine paternal identity (e.g., elephants, chimpanzees). In situations where food is scattered and limited, both parents need to provision the young.

POPULATION GENETICS AND BEHAVIOUR

Population genetics described the statistical distribution of genes in a particular breeding population, such as a breed of dog, and examines the role of various biological effects on gene distribution. While ordinary genetics predicts phenotypic makups of the next generation, population genetics concerns itself with genetics predictions for a breeding group several generations into the future.

A population is the collection of organisms of a particular species within a given geographic area. When all individuals in a population are identical with regard to a particular phenotypic trait they are known as monomorphic. When the individuals show several variants of a particular trait they are polymorphic. The Gene pool for a species or population is the complete set of unique alleles present in it. While large gene pools indicate robust populations, extensive genetic diversity, and a good chance to survive bouts of intense selection, low genetic diversity is associated with reduced biological fitness and an increased chance of extinction.

Hardy-Weinberg Equilibrium

Allele frequencies will remain constant across generations if the population is large and individuals are mating randomly. Moreover, we will be able to acurately predict genotype frequencies from gene frequencies (e.g., in a two-allele system—a2+2aA+A2). HW produces the null hypothesis against which genotype frequencies are compared.

HW equilibrium will occur after one generation provided that the same gene frequencies occur in both sexes. If disequilibrium occurs, equilibrium will be re-established after one generation of random mating. What are the assumptions? What will produce deviations from the HW equilibrium? What might produce an excess of heterozygotes (e.g. HLA-1, MHC)? What might produce an excess of homozygotes? Deviations from HW equilibrium mayt derive from a variety of sources.

Mutations

Mutations are alterations in genetic code ranging from changes to individual nucleotide bases along the DNA to largescale rearrangements of chromosomes. Mutations are rare and even at maximum rates allele frequencies will only change from p=1 to p=0.9993 per generation.

Genetic Drift

With small effective population size, random events can take on increasingly prominent roles. This focuses on changes in the genetics of the population as a result of chance alone.

— Bottlenecks

— Founder Effect

Examples include the relatively high rate of an autosomal recessive skeletal dysplasia (Ellis-van Creveld Syndrome) in the Amish population of Lancaster, Pa. Resulting in short-limbed disproportionate dwarfism, its incidence within the population is approximately 5 per 1000 live births and 2 per 1000 living persons. It can be traced to the initial immigrants Samuel King and his wife.

Selection

Strong selection for traits with high heritability can change proportions in excess of 10% per generation

Non-random Mating

Mate Choice with assortative mating, Positive assortative mating: mating among individuals who share particular genes or phenotypes. Negative assortative mating: preferrence for mating among individuals who are unlike each other with respect to particular genes or phenotypes.

Inbreeding and Outbreeding. Inbreeding: refers to (preferential) mating between biological relatives. As relatives (i.e., ancestors of the first individuals are shared with those of the second individual), they carry genes which are "identical by descent"; extreme inbreeding: mating between sibs, half-sibs, parent-offspring. Individual inbreeding coefficient (i.e., Pedigree inbreeding): F represents the probability that the offspring is homozygous due to identity by descent (ibd) at a randomly chosen autosomal locus, ranges in value from 0 (no locus ibd) to 1 (all loci ibd). Significant factors for pedigfree inbreeding are

Factors which determine the probability that various kinds of individuals will come into contact, e.g., population demography (sex ratio, birth or death rates), individual and population dispersal patterns, spatial distributions Behavioural preferences for or against certain classes of individuals as potential mates. Behavioural preferences influence the acceptance or rejection of different types of individuals as mating partners

— *Population inbreeding coefficient*: f measured by deviations from Hardy-Weinberg Equilibrium (a2+2aA+A2), examine for an over-representation of homozygotes. f ranges from -1 (inbreeding avoidance) to 0 (random mating) to 1 (inbreeding)

— *Outbreeding*: (preferential) mating between non-relatives

— *Inbreeding depression*: overrepresentation of homozygotes; e.g. matings between cousin are rare (0.05% in U.S.) but account for 20% of albinos; deficiency in immune systems; recessive, deleterious alleles occur more likely in homozygous condition; humans have on average 8 recessive, deleterious alleles. 1% increase in inbreeding produces 11% increase in the incidence of genetic deseases; Even small amounts of pedigree inbreeding in a random mating population greatly increase the incidence of some types of genetic disease. This is of particular concern in groups with small effective population sizes—Amish, Ashkenzi Jews.

— *Inbreeding advantages*: functional association of gene complexes shaped by local adaptations

— *Outbreeding depression*: Extreme case, hybridisation across genetically differentiated populations

— Hybrid vigor

ALTRUISM AND SOCIALITY

Many animals and humans show a startling flair for exhibiting kindness, cooperation, and altruistic behaviour. Evolution, which is often (and falsely) viewed as largely driven by conflict between competing individuals, may not be an obvious candidates for explaining their existence. Evolution may promote traits underlying group living, altruism, and coorperative behaviours but only if genes underlying such characteristics are favored.

— Cooperation refers to the practice of individuals working together instead of working separately in competition. The success of participating individuals depends and is contingent upon the success of their partners in cooperation.

— Altruism consists of unselfish/selfless acts in which the altruist appears to put other individual's interests before their own. The general expectation is that the evolution of altruistic traits is not compatible with selfish gene theory. How could such a mechanism reinforce a behaviour jeopardising one's own interests at the expense of another's? Examples include self-sacrificing Behaviour: e.g. vampire bats, guard bees, Cooperative breeding, e.g. African Kingfishers, Acorn Woodpeckers, Florida Scrub Jay;

— Social behaviour relates to interactions with other members of the population. Group living comes with a diverse range of advantages and disadvantages. A Society describes a group of individuals of the same species, organised in a cooperative manner, and extending beyond sexual contexts and parental care.

— Eusociality is an extreme form of altruism where sterile workers labor on behalf of reproductive individuals, cooperate in the care of young. There must be an overlap of at least two generations of individuals which share in colony labor.

Units of Selection

Discussion about the units of selection centers on an understanding of what the life of each individual is ultimately devoted to. This is nowhere more pertinent than in evolutionary explanations for altruistic acts or for the emergence of highly social societies.

— *Group Selection*: Altruistic behaviour could be rooted through group selectionist thinking where altruistic traits favor reproductive success of a group at the expense of that of an individual (Wynne-Edwards). Groups containing altruists may thereby be able to outcompeting other groups, which lack such cooperation. Unfortunately, all members of a group displaying a high level of cooperation will derive the benefits, even those individuals who may be less cooperative than others. "For the good of the species..." explanations have not fared well in theoretical or empirical studies.

— *Individual Selection*: If individuals are simply looking out for their selfish interests, how could altruistic behaviour ever emerge as a superior strategy to it?

— *Gene Selection*: Is altruistic behaviour focused on the preservation of selfish genes

Simple Egoistic Solutions

In some instances, benefits may accrue to an individual committing an altruistic act. If these benefits exceed its cost, it is in the egoistic interest of the individual to perform such an altruistic act.

Synergistic Benefits

Cooperation may be in each individual's egoistic interest if it increases its own individual fitness. Wolves, for instance, hunt in packs and then share their prey. Thereby each wolve benefits from pack hunting as the group can bring down prey that would be too large for a single individual. Although the behaviour may seem altruistic, individuals benefit directly in tasks that require many eyes. Fish tend to group in schools as it becomes significantly harder for a predator to pick out and hunt one specific individual from within their midst. In 1971, Hamilton proposed the Selfish Herd model, which postulated a centripetal instinct that directs animals to place themselves towards the centre of a group. This behaviour is based on a presumed increase in predation towards the outer edges of the group.

Delayed Benefits

Young may remain with their parents and help them or group members raise more offspring, rather than breeding themselves (e.g., ostriches, some primates). Helpers may gain useful experience in raising their own offspring or have an increased chance to inherit a valuable breeding territory.

Reciprocity

Reciprocal altruism where the benefits accrued to the individual exceed the costs it incurs Kin selection when behaviour sufficiently enhances the fitness of related individuals (i.e., carrying similar genes) leading to an increased representation of genes being passed on to the next generation.

True Altruism

An individual behaves in such a way as to enhance the reproduction of another individual, at a cost to its own fitness. (e.g., Sterile workers in social insects who give up all reproduction for the benefit of their mother queen. Mutual or delayed benefits can't account for this one: sterile workers never get to produce any daughters.)

Inclusive Fitness

An individual fitness solution to the paradox goes as follows: genes are favored that produce a disproportionate propagation of ones genes to subsequent generations even if it is at the expense of the individual. This may occurr directly through either reproducing personally or indirectly by encouraging the reproduction of close relatives who share many of ones genes. Under those circumstances altruistic behaviours should primarily be directed towards close relatives who share many of the genes as compared to non-relatives who share fewer of them.Inclusive fitness refers to the sum of an individual's *Direct Fitness* (probability of reproductive success of one's own offspring) and its *Indirect Fitness* (probability of reproductive success of non-descendant relatives). Altruistic acts are frequently directed towards relatives

Kin Selection: Selection for traits that lower an individual's personal fitness, but raise a relative's fitness. Recipient (related) kin share genes with the altruistic individual and are thus genetic extensions of them. Kin selection and inclusive fitness can explain sociality through increased survivorship of relatives, increased inclusive fitness, altruism, and delayed maturation or breeding.

Inclusive Fitness: refers to the sum of an individual's Direct Fitness (probability of reproductive success of one's own offspring) and its Indirect Fitness (probability of reproductive success of non-descendant relatives). Altruistic acts are frequently directed towards relatives. <Kin Selection>: selection for traits that lower an individual's personal fitness, but raise a relative's fitness. Recipient (related) kin share genes with the altruistic individual and are thus genetic extensions of them. Kin selection and inclusive fitness can explain sociality through increased survivorship of relatives, increased inclusive fitness, altruism, and delayed maturation or breeding.

The complx needs for group-living may have driven members towards a surprising degree of intelligence. Individuals of many species are able to keep track of each other's social status, correctly attribute each others mental states, and successfully engage in each other's deception.

MODELING ADAPTIVE BEHAVIOUR

It is a daunting (or impossible) task to study behaviour by simultaneously considering all significant factors and interactions that may impact it. An alternative method attempts to reduce the associated complexities by wrapping ones exptectations into a set of mathematical abstractions based on a reduced number of significant variables.

Resulting quantitative predictions are necessarily incomplete but are often less ambiguous and easier to test. The goal is to model an animals rational decisions of what action to take, given some information about the world. The player then faces consequences for the decision as a function of the action, the actions of others (if applicable), and the state of the surrounding environment. The player is expected to be rational by maximising the expected utility.

Optimality Models

Optimality models quantitatively predict the consequences for a particular animal to behave in a certain way when pitted against the environment. It assumes that animals should behave so as to maximise their fitness. Optimality models consider separate, dependent, fitness benefits (B) and fitness costs (C) over a given range of values in a decision variable (i.e., the variable of interest). The solution to an optimality model attempts to find the point where the net benefits (i.e., B-C) are maximised. Individual solutions consider the consequences for a focal animal regardless of what strategies other individuals are relying on.

The main advantages of optimality models are in that they:

- help us clarify our assumptions
- force us to think about the information the animal has
- tie many different ideas together in one place
- generate testable predictions

Consequences of behavioural decisions are described as equations where the success of decisions of one individual do not depend on what decisions other animals make.

- Decision variable is the variable that we aim to optimise (e.g., optimise speed to obtain the best mileage when driving from here to there)
- Currency refers to the criterion used to examine the outcome for different values of the decision variable (mileage)
- Constraints are factors that limit the relationship between decision variable and currency (e.g., speed limits)

Optimality models use calculus and logic to solve for a minimum or a maximum in a specific function f(x). This can be achieved by taking the derivative of the function df[x)/dt (i.e., to obtain the slope of the function) and setting it to 0. Such an approach can also be used to explore sensitivity to changes in values of the decision variable and to determine what types of information are relevant for the relationship. Precise predictions can be formalised for empirical testing

Optimal Foraging Theory

OFT examines the choice of food items. Such prey models (even though the prey doesn't necessarily have to be another animal) look at decisions of basic diet selection. The animal searches for food, and finds potential prey items one at a time. It has to decided whether it should stop searching and eat the food it has found, or whether to ignore the food and keep searching.

We assume that the animal is trying to maximise its currency (i.e., rate of energy intake measured in calories) in a particular time period. Each type of food takes a different amount of time to handle it during consumption. For instance, Northwestern Crows search along the water line during low tide for large molluscs called whelks. When it finds one, it flies upward and drops it on the rocks to break it open. Smaller whelks are harder to break, and need to be dropped from a greater hight, so they may be ignored. Also the content of certain kinds of nuts is more difficult to access than that of others.

E1 and E2 refer to the caloric energy that is contained within a prey of type 1 or 2 respectively. The handling time needed to access the caloric content for such prey items is h1 and h2. If the net energy gained per unit of time (or the rate of energy gain) for type 1 exceeds that for item 2

$$E_1/h_1 > E_2/h_2$$

then you would do well to eat prey of type 1 whenever you find it. If you find an item of type 2 you may consider rejecting it and to keep searching if the energy lost in searching for item 1 added to its handling time is less than what is gained from eating item 2 outright.

$$E_2/h_2 > E1/(S_1 + h_1)$$

Depending on the relationships of this equation the predator should eat just prey 1 (and be a specialist) or it should eat both prey 1 and prey 2 (and be a generalist). This decision to specialise depends on the quality and abundance of S_1, regardless of S_2.

Marginal value theorem

The MVT considers an optimally foraging creature that exploits patchy resources as it must decide when to move on to the next patch. As it attempts to optimise a cost/benefit ratio individuals will stay longer as the distance between patches increases or when the environment as a whole is less profitable.

Game Theory Models

A more comprehensive prediction of an animal's most profitable behaviour may well require us to also consider what others are doing. A behaviour may well be quite rewarding when it is rare in a population (e.g., deception) but may not be nearly as advantageous to its actor when it becomes common.

Game theory models thus consider the value of frequency-dependent fitness benefits. Originally developed as a tool to predict rational human economic behaviour, its application to many evolutionary problems has improved our understanding of situations where fitness consequences of a behaviour depend on types and frequencies of behaviours exhibited by other animals within the population.

It is assumed that individuals represent players of a game in which all parties are aware of the rules, are consciously attempting to maximise their payoffs, and are attempting to predict the moves of their opponents. Players move simultaneously or at least do not observe the other player's move before making their own.

A Pay-off matrix for a hypothetical, normal form game that pits two individuals against each other. Player Blue (rows) may chose strategy A or B when dealing with animal 2, while player Red (columns) may rely on strategy 1 or 2. The payoff matrix lists the colourcoded consequences that each individual receives for the combination of strategies played.

	strategy !	strategy 2
Strategy A	3, 3	2, -1
Strategy B	-1, 2	2, 2

A Payoff Matrix is used to formally represent a game that includes all conceivable strategies, along with each strategie's corresponding payoff against any other strategy. The matrix of a game specified the payoffs that each player receives for the combinations of actions played. For example, if player 1 plays top and player 2 plays left, player 1 receives 4 and player 2 receives 3. In each cell, the first number represents the payoff to the row player (in this case player 1), and the second number represents the payoff to the column player (in this case player 2).

A case of a symmetric game exists if the payoffs do not depend on which player chooses each action. In the example above, this is the case if strategy A (player 1) is the same as strategy 1 (player 2) and strategy B of player 1 is the same as strategy 2 of player 2. The game can then be represented with a single payoff per cell, i.e., the payoff for the row player. For example, the payoff matrices on the right represents the same game as the matrix above, if 1 = A and 2 = B..

	strategy A	strategy B
strategy A	3	2
strategy B	-1	2

The simplified Pay-off matrix rewritten for a hypothetical, symmetrical, normal form game that pits two individuals against each other. As both player may chose strategy A or B, the payoff matrix lists the consequences that the row individual receives for all possible combinations.

Game theory aims to identify behavioural rules that form an Evolutionary stable strategy (or ESS). This strategy, if adopted by a population of players, cannot be invaded by any mutant strategy. An ESS is "evolutionarily" stable meaning that once it is fixed in a population, natural selection alone is sufficient to prevent alternative strategies from invading the population.

Prisoner's Dilemma

Hawks and Doves Game

Game models have greatly increased our understanding of conditions under which ritualised displays may take the place of aggressive interactions. Maynard-Smith's approach involves resource-centered conflict between members of a group which will follow either a strategy of unrestrained use of weapons (hawk) or reliance on highly ritualised elements of fighting (dove):

	Hawk	Dove
Hawk	(V-C)/2	V
Dove	0	V/2

Payoff matrix for the Hawk-Dove game. Hawk is a strategy where the individual fights until either it sustains an injury or the opponent retreats. The losing hawks suffers a cost for such injury (C). A Dove contests the interaction with displays but will retreat immediately if the opponent escalates with a use of its weapons. The winner of the encounter obtains the resource in question and gains control over its value (V). Interactions between two hawks result in the term 1/2(V-C) as hawks against hawks will win half of the time and lose half of the time. Moreover, hawks will also sustain an injury half of the time. In interactions with another dove, a dove will win half of the time and retreat without cost the other half of the time.

The different cells of the payoff matrix represent combinations of strategies for the two players, where each would prefer to win, prefer to tie rather than lose, and prefer to lose over injury. Game theory can be used to address the conditions under which such strategies are evolutionarily stable (i.e., an ESS). First we ask if a population playing one strategy can be invaded by a few animals playing the other strategy. To examine whether dove is an ESS we examine whether a population of doves can be invaded by a hawk?

$$u[H,D] > u[D,D]$$

$$V > V/2$$

A hawk can invade a population of doves as its payoff (V) is greater than that of a dove pitted against other doves (V/2). Dove thus cannot be an ESS. Hawks will be able to invade a population of doves until encountering another hawk becomes a common event. To examine whether a dove can invade a population of hawks, we test whether its payoff is greater than that of a hawk when faced with a population of hawks. Formally this evaluates whether

$$u[D,H] > u[H,H]$$
$$0 > (V-C)/2$$
$$0 > V/2-C/2$$
$$0 > V-C$$

The conclusion on whether a dove will be able to invade a population of hawks will thus depend on whether V > C. If the resource value exceeds the costof injury, the benefit to hawks will be positive (i.e., greater than 0), and a population of hawks cannot be invaded. Hawk thus represents an ESS under such conditions. However if V<C then payoff to hawks will be negative and doves can invade.

Doves become increasingly good at invading a population of hawks as opportunities for damage increase. In fact a population of hawks would go extinct in situations where the cost of injury exceeds the resource's value. To obtain the proportion of hawk and dove strategies that stably coexist when C>V we examine at what proportion of Hawks (p) and Doves (1-p) the fitness of individuals playing either of the two strategies is equal

$$W[H] = p*((V-C)/2) + (1-p)*(V) \text{ equals } W[D] = p*0 + (1-p)*(V/2)$$
$$p*(V/2-C/2) + (1-p)*(V) = (1-p)*(V/2)$$
$$p(V/2)-p(C/2) + V-pV = (V/2)-p(V/2)$$
$$p(V/2) + p(V/2)-pV-p(C/2) = (V/2)-V$$
$$- p(C/2) = -(V/2)$$
$$pC = V$$
$$p = V/C$$

Game theory thus provides a theoretical underpinning for why animals tend to only fight with great ferocity when a resource of great value is at stake. Fighting, for instance, is particularly intense in elephant seals where victorious males are able to monopolise a large area of beach which contains many females. In the great majority of instances, however, resources are rarely worth being injured over and competing individuals will resolve most conflict with ritualised displays.

Chicken, Brinkmanship, and War of Attrition Games

The game "Chicken" is a variant of the Hawks and Doves game. In it each player prefers not to yield to the other, but the outcome where neither player yields is by far the worst possible one for both players. It has its origins in the situation where two drivers head towards each other on a collision course: one must swerve, or both may die in the crash. The driver who does swerve is called a "chicken".

Payoff matrix for a game of chicken.

	swerve	straight
swerve	0	-1
straight	1	-10

The game's most important feature is an assured, and significant cost to the situation where both individuals chose conflict over conciliation. "Losing" by deciding to swerve is so trivial compared to the crash that occurs if nobody swerves, the reasonable strategy would be to swerve. Yet, knowing this one may well decide not to swerve, gambling that the opponent will be reasonable.

Insights gained on the predictability of behaviour based on the presence of overwhelming punitive consequences have formed a central component of the cold war military doctrine of Mutual assured destruction, where the use of nuclear weapons by either side would assure the destruction of the aggressor along with its target. This theory of deterrence states that the deployment of strong weapons as a threat to the enemy is essential in order to prevent the use of the very same weapons. Success in discouraging either side from resorting to these weapons is only effective as it is credible to both sides that a chosing conflict would result in the worst possible outcome for all—nuclear annihilation of both sides.

Fighting in juvenile American lobsters (*Homarus americanus*) begins with a series of threat displays. If opponents are evenly matched, the encounter progressively escalates through ritualised components of fighting, restrained forms of physical combat, and finally brief periods of unrestrained fighting where opponents may even inflict injuries on each other.

War of attrition scenarios refer to the situation where fighters attempt to grind down the opponent's defenses. There is no fixed cost associated with losing or contesting, but as the encounter wears on, each player accumulates incremental costs. A decision to give up arises when one indidivual backs down, relinquishing access to the contested resource, rather than continuing to sustain further insults.

EVOLUTIONARY ENVIRONMENT AND HUMAN BEHAVIOUR

In both humans and other animal species, evolved behaviour patterns reflect the selective pressures of the ancestral environments. Evolutionary psychologists, following John Bowlby, sometimes speak of the EEA or environment of evolutionary adaptedness. This is the environment in which an evolved tendency was adaptive. In other words, it is the environment in which a behaviour evolved. A simpler term is ancestral environment, if the term is understood to refer to the period of time when a behaviour evolved, not earlier or later times.

For humans, the EEA was the environment of the most recent 125,000 years or so. Lactose tolerance, the example used above, evolved only recently, in the past 2,000 years. That is extremely recent by evolutionary standards, and it shows the great advantage milk-drinking provided to people in northern climates. Many human behaviours presumably evolved during a much earlier period of pre-history.

DNA studies show that biologically modern humans emerged from a small group of approximately 1,000 individuals who lived about 100,000 years ago. The population bottleneck (ours was an endangered species at the time) made the human race genetically uniform.

Geneticists say there is less variation between the DNA of humans from opposite ends of the earth than there is between gorillas or chimps from adjacent patches of jungle. The reason is that gorillas and chimpanzees experienced no population bottleneck in their recent evolutionary past, so their populations have had more time to accumulate random changes and variations. Humans, by contrast, are all very similar in their DNA.

Organised agriculture began relatively recently, about 10,000 to 20,000 years ago. After that, many groups of modern humans lived in villages. However, cultural change occurs much faster than biological change. Most human behavioural characteristics were already in place by the time humans learned to domesticate plants and animals. Most evolutionary psychologists believe the dominant social environment for evolving humans was the older, hunting-and-gathering, nomadic way of life.

The adaptive value of a behaviour, or any other trait, is always relative to some environment. When the environment changes, so does the adaptive value of a particular behaviour. On the famous Galapagos Islands where Darwin made many of his observations, the Woodpecker Finch uses twigs and cactus spines to probe for insects in holes of dead trees. This is an excellent example of an adaptive behaviour in a particular environment. It represents an adjustment to the environment of the islands, where insects in the trees provide food for the finches, which gives the finches an evolutionary advantage. By exploiting this food source, they are more likely to survive and reproduce than finches that cannot use twigs in this way.

However, the finch's behaviour is adaptive only because there are no woodpeckers on the island. If there were woodpeckers in the area, the finch's behaviour would no longer give it a reproductive advantage. Woodpeckers are more efficient at extracting insects from dead wood, so food would not be available for the twig-using finch, and it might become extinct.

To interpret the fitness or adaptive value of a behaviour, one must specify the environment. That includes the other creatures present. The introduction or extinction of a single new species can tip the balance of survival for dozens of other species in that ecosystem. For exapmle, cats introduced to island ecosystems by humans have a devastating effect. They are excellent predators and the animals living on isolated islands have no evolved defenses against them.

Fire ants have devastated competing ant species in the southern U.S. In Europe, the arrival of modern humans around 40,000 years ago probably caused the extinction of the Neanderthals, who had lived there for 300,000 years. Again and again, the fossil record shows that the arrival of modern humans in an ecosystem is followed by the mass extinction of large animals in the same area, probably due to hunting. Early humans are now believed to have hunted the wooly mammoth to extinction.

Humans are often so successful at modifying their environments that they bring about the destruction of their own societies. The Maya civilisation burned wood to cook limestone, which was ground up and used to make a form of concrete, the main construction material for their massive ceremonial temples. Simple calculations show that this consumption of forest resources was unsustainable. Eventually all the trees were gone, and that (along with a drought and a belief system predicting the end of the world) led to the end of Mayan civilisation.

A similar process occurred on Easter Island, where an isolated population of humans cut down the native species of trees until they were all gone. The ecosystem was devastated and the humans eventually faced repeated famines. Many similar events are occurring today, for example, in fishing villages of the northeastern United States, where a traditional way of life is dying because the level of fishing has reduced stocks to the point where people can no longer make a living by fishing.

On issue after issue—including some of the most significant problems of modern times—the challenge facing our species is to modify our behaviour intelligently so we can encourage cultural and behavioural practices that are healthy and sustainable, while discontinuing those that are self-defeating. Evolutionary psychology may help. In some cases, simulations may suggest adaptive strategies for the future. In many other cases, evolutionary psychology can help by explaining the origins of destructive behaviour patterns left over from the past. If we understand where they came from and how they might have been adaptive in ancient times, it may help us to move beyond them and not to continue blindly with practices that are maladaptive in the modern world.

REFERENCES

Alcock, J. 2001. *Animal Behaviour: An Evolutionary Approach* (7 ed.) Sinauer Associates Inc., Massachusetts.

Hamilton, WD. 1964. "The genetical evolution of social behaviour I and II". *J. Theoretical Biology* 7: 1-16 and 17-52.

Williams GC. 1966, *Adaptation and natural selection: a critique of some current evolutionary thought*. Princeton, N.J: Princeton University Press.

Wilson EO. 1979, *On Human Nature*, Harvard University Press.

3

Animal Behavioural Genetics

Behavioural genetics is the field of biology that studies the heritability of behavioural traits in animal (including human) behaviour. The field has formed from an overlap of genetics, ethology and psychology. Genetic characteristics are those that are to a large extent determined by genes. Although genes may play a role in many behaviours, they never determine them.

There are no genes that directly code for a behaviour—genes only code for proteins. However, it is clear that a change in a single protein can cause a host of downstream effects and may even bring about a distinct phenotype.The external environment exerts a strong influence on how all genes are expressed in behaviour via a development of nervous and hormonal mechanisms. The Phenotype (i.e., the observable characteristics of an organism) emerges from an interaction of its Genotype (i.e., the organism's genetic composition) with environmental factors.

A Gene is the smallest functional unit of heredity and is composed of DNA. It specifies the codes for amino acid chains that make up individual proteins (e.g., serum albumin). A Locus is a section on a chromosome that relates in a meaningful way to a function. Monomorphic loci are loci with a single common segment of code, and which is present in 95-99% of individuals. In contrast, polymorphic loci exist in form of two or more alleles (i.e., versions of the same position of the chromosome) with a combined frequency>0.05. Diploid organisms carry two copies of each gene, one derived from each parent. Homozygous is the condition in which both of these copies for that gene are identical (i.e., the same allele). If the chromosomes contain two different alleles of the gene, it is called heterozygous. In some cases the phenotype of heterozygous individuals follows one allele (i.e., the dominant one) and not the other (i.e., the recessive allele).

Recessive phenotypes will only be expressed in cases where the individual is homozygous for that allele. If one gene pattern is well established and common in a particular organism, it is referred to as the wild type allele. In contrast, a mutant allele usually represents a relatively new, and less common modification. Pleiotropy refers to the situation where there is no clear 1:1 correspondence between one particular gene and a particular behavioural trait. For instance, fruitflies with a mutant *period* gene display a range of behavioural effects, ranging from differences in daily activity rhythms, time of day at which adults hatch from pupae, and the particular patterning of the male's courtship song.

Heritability is a statistical estimate of variation in a trait that is attributable to genetic differences among individuals within a group. For instance, the number of arms humans have is heritable and can be expressed as a proportion. Variables such as an individual's fingerprint and height are strongly heritable, while other variables such as musical abilities are much less so.

CLASSICAL GENETICS

Classical (or Mendelian) genetics examines the distribution of hereditary characteristics of behaviours from one generation to the next. With selective breeding, the presence/absence of specific behaviours can be tracked through the outcomes of sexual reproduction. From such work came the realisation that genes existed, that individuals are diploid with one half of the genes contributed by each parents, and that some genes dominate the phenotype over others. The genetics of many diseases follows simple Mendelian rules while examples of behaviour in this category are generally few. Rather than in a small number of discrete states, most morphological/behavioural traits (body size, intelligence, boldness) occur across a continuous phenotypic range.

Seymour Benzer discovered a series of behavioural mutants in fruit flies, including single gene disruptions of circadian rhythms (*period*), and various genetic causes in neurodegeneration and aging. Hygienic behaviour in honey bees involves the ability to detect and remove diseased, larval and pupal brood from the nest before the pathogen becomes infectious. Forming a mechanism of resistance to bee diseases and parasitic mites, it consists of two distinct task-components: uncapping a cells and removing its content. Rothenbuhler suggested that these two traits were controlled in a simple Mendelian manner by two recessive loci. More recent molecular evidence from quantitative trait loci (QTL) linkage mapping has identified multiple specific stretches of DNA with genes that underly variation in this trait. This work suggests that the genetic basis of hygienic behaviour is considerably more complex, and that seven QTLs are associated with hygienic behaviour, each controlling only 9-15% of the observed phenotypic variance.

QUANTITATIVE GENETICS

Quantitative genetics explores the inheritance of characteristics that. Such traits are often subject to significant environmental influences and usually involve a great number of underlying genes. Although each gene may well follow a simple Mendelian pattern, overall results are not easily characterised in these terms. Quantitative approaches are essential to studying the degree to which relatedness among Individuals is matched by resemblances in behavioural characteristics. It thus aims to estimate the proportion of total phenotypic variance that is explained by relatedness. Aside from cases where multiple genes affect single traits, changes in one individual gene may impact quite a number of different traits. Such instances can be identified and characterised through measures of phenotypic covariance across multiple traits.

POPULATION GENETICS

Population genetics concerns itself with mechanisms that change the relative occurrence of genes within a population. It examines how gene frequencies change or become stabilised within populations through behavioural forces such as sexual selection, mating systems, dominance, or territoriality.

DEVELOPMENTAL GENETICS

Developmental genetics explores how genetics interfaces with ontogenetic processes in behaviour. Genetics plays an integral role in the control of cell growth and differentiation, formation of tissues, organs and hormone systems, as well as in the emergence and critical timing of learning opportunities, cognitive abilities, and emotional systems. Microarray technology is able to characterise levels of activity for many genes simultaneously. This technique measures how many copies of different mRNAs are made when genes are turned on.

GENETIC EFFECTS OF DOMESTIFICATION

Domestication is a process by which a population of animals becomes adapted to man and the captive environment by some combination of genetic changes occurring over generations and environmentally induced developmental events recurring during each generation:' In long-term selection experiments designed to study the consequences of selection for the tame" domesticated type of behaviour, Belyaev and Belyaev *et al.* studied foxes reared for their fur.

The red fox (*Vulpes fulva*) has been raised on seminatural fur farms for over 100 years and was selected for fur traits and not behavioural traits. However, they demonstrate three distinctly different characteristic responses to man. Thirty percent were extremely aggressive toward man, 60% were either fearful or fearfully aggressive, and 10% displayed a quiet exploratory reaction without either fear or aggression.

Figure 1. Vulpes fulva

The objective of this experiment was to breed animals similar in behaviour to the domestic dog. By selecting and breeding the tamest individuals, 20 years later the experiment succeeded in turning wild foxes into tame, border collie-like fox-dogs. The highly selected "tame" population of (fox-dog) foxes actively sought human contact and would whine and wag their tails when people approached. This behaviour was in sharp contrast to wild foxes which showed extremely aggressive and fearful behaviour toward man. Keeler *et al.* described this behaviour:

Vulpes fulva (the wild fox) is a bundle of jangled nerves. When first brought into captivity as an adult, the red fox displays a number of symptoms that are in many ways similar to those observed in psychosis. They resemble a wide variety of phobias, especially fear of open spaces, movement, white objects, sounds, eyes or lenses, large objects, and man, and they exhibit panic, anxiety, fear, apprehension and a deep trust of the environment. They are

1) Catalepsy-like frozen positions, accompanied by blank stares;
2) Fear of sitting down;
3) Withdrawal;
4) Runaway flight reactions; and
5) Aggressiveness.

Sometimes the strain of captivity makes them deeply disturbed and confused, or may produce a depression- like state. Extreme excitation and restlessness may also be observed in some individuals in response to many changes in the physical environment. Although the stress of domestication is great, Belyaev and Belyaev *et al.* concluded that selection for tameness was effective in spite of the many undesirable characteristics associated with tameness. For example, the tame foxes shed during the wrong season and developed black and white patterned fur, and changes were found in their hormone profiles. This means that the monoestrus (once a year) cycle of reproduction was disturbed and the animals would breed at any time of the year. Furthermore, changes in behaviour occurred simultaneously with changes in tail position and ear shape, and the appearance of a white muzzle, forehead blaze, and white shoulder hair. The white colour pattern on the head is similar to many domestic animals.

The most dog-like foxes had white spots and patterns on their heads, drooping ears, and curled tails and looked more like dogs than the foxes that avoided people. The behavioural and morphological (appearance) changes were also correlated with corresponding changes in the levels of gender hormones. The tame foxes had higher levels of the neurotransmitter serotonin. Serotonin is known to inhibit some kinds of aggression, and serotonin levels are increased in the brains of people who take Prozac (fluoxetine).

The study of behavioural genetics can help explain why selection for calm temperament was linked to physical and neurochemical changes in Belyaev's foxes. Behaviour geneticists and animal scientists are interested in understanding effects on behaviour due to genetic influences or those which are due to environment and learning.

HISTORY OF BEHAVIOURAL GENETICS

Early in the 17th century, Descartes came to the conclusion "that the bodies of animals and men act wholly like machines and move in accordance with purely mechanical laws". After Descartes, others undertook the task of explaining behaviour as reactions to purely physical, chemical, or mechanical events. For the next three centuries scientific thought on behaviour oscillated between a mechanistic view that animals are 'automatons" moving through life without consciousness or self-awareness and an opposing view that animals had thoughts and feelings similar to those of humans.

In "On the Origin of the Species", Darwin's ideas about evolution began to raise serious doubts about the mechanistic view of animal behaviour. He noticed that animals share many physical characteristics and was one of the first to discuss variation within a species, both in their behaviour and in their physical appearance.

Darwin believed that artificial selection and natural selection were intimately associated and cleverly outlined the theory of evolution without any knowledge of genetics. In "The Descent of Man" Darwin concluded that temperament traits in domestic animals are inherited. He also believed, as did many other scientists of his tune, that animals have subjective sensations and could think. Darwin wrote: "The differences in mind between man and the higher animals, great as it is, is certainly one of degree and not of kind."

Other scientists realised the implications of Darwin's theory on animal behaviour and conducted experiments investigating instinct. Herrick observed the behaviour of wild birds in order to determine, first, how their instincts are modified by their ability' to learn, and second, the degree of intelligence they attain.

On the issue of thinking in animals, Schroeder concluded: The solution, if it ever comes, can scarcely fail to illuminate, if not the animal mind, at least that of man." It is evident that by the end of the 19th century, scientists who studied animal behaviour in natural environments learned that the mechanical approach could not explain all behaviour.

Gregor Mendel, the father of modern genetics, found single genes responsible for producing individual traits in his landmark experiments with peas in the 19th century. He noticed that his peas' phenotypes, or actual appearances, were dictated by their genotypes, or genetic makeup. While Mendel could develop a simple model of inheritance of single traits for the all the peas in his garden, human genetics is complicated by the fact that many traits we possess are dependent upon the actions of many genes as well as environmental factors. The traits Mendel studied in his peas facilitated genetic analysis because they fell into distinct categories such as dwarf or tall, yellow or green. Most human traits cannot be described in such a simple manner, but instead exist as a continuous distribution of characteristics. Height is one such trait. Human behavior is another, but is even more complex than height because it cannot be quantified as readily. The difficulty in determining a method for measuring behavior makes the problem of defining the relationship between genotypes and behavioral phenotypes extremely challenging.

Sir Francis Galton, inspired by his cousin Darwin's Origin of Species (1859), pioneered the field of behavioral genetics, developing a school of thought called hereditarianism. Galton believed all human traits, including behavior, are solely determined by genes, without any contribution from the environment. This idea became the basis for eugenics, a term Galton coined in 1883 to describe the use of genetics for social planning. Galton, in other words, believed that genotype alone determined most characteristics, and that selective "breeding" of the human species could guarantee that the "best" traits would remain in the human population while the "worst" traits could be eliminated.

Based on Galton's ideas, the eugenics movement built momentum in the United States, where between 1905 and 1933, several states based laws and legal decisions on eugenics theory, and the federal government heavily restricted immigration from Eastern and Southern Europe. The goal was the production of a genetically "pure" American population. During this time in American history, criminals, people with low IQs and, in some cases, even women who had illegitimate children, were sterilised.

During its time in power, the government of Nazi Germany staunchly supported and enforced the principles of eugenics. Eugenics-based legislation culminated in the "euthanasia" of the mentally retarded and the physically disabled, who, according to Nazi geneticists, would otherwise contaminate the German people with unfit genes. The list of genetic undesirables ultimately grew to include Jews, Gypsies, Communists, Catholics, homosexuals, and political opponents of the government. During the 1930s and 1940s, a large portion of the state's resources funded a system of death camps which, by the end of World War II, exterminated nearly twelve million people.

Galton's association with the eugenics movement often overshadows the fact that he made major contributions to the study of human genetics and statistical analysis of traits. It was Galton who first suggested some of the major methods of human behavioral genetics used today. He conducted the first systematic family, twin, and adoption studies to show how behavioral traits can run in families. In addition, he devised the concept of correlation, an important tool used to describe resemblances among family members and a fundamental concept of statistical analysis today.

Largely tainted by the ideas of the eugenics movement, twin and adoption studies conducted during the 1920s and 1930s played only a minor role in the advancement of behavioral genetics. During the same period, studies demonstrating the importance of genetics on the behavior of animals actually had greater impact on the development of today's behavioral genetics. Scientists carried out selection experiments to create breeds of rats, mice, and dogs characterised by particular behavioral traits such as curiosity and emotionality.

In the 1960s, studies on the genetic influence of traits such as IQ once again gained notoriety, but their results quite often remained ignored because of their associations with eugenics and racism. It was not until the 1980s that the scientific community began to pay serious attention to behavioral genetics as a science. Using gene manipulation technologies which had been in the works since the late 1960s, geneticists could finally map the human genome itself.

Since the early '80s, geneticists have identified genes for cystic fibrosis, muscular dystrophy, and Huntington's disease. The successes in identifying and mapping human disease genes have revitalised interest in identifying genetic factors underlying behavioral traits. Twin and family studies, genetic studies of animal behavior, biochemical investigations of mutations, and the development of new molecular approaches, have all added to the momentum of the modern search for genes influencing behavior.

During the Bush administration, the U.S. government announced a plan to begin a "violence initiative" to find the causes-genetic and otherwise-of violence. While some studies did receive funding, the initiative as a federal programme was scuttled at the last minute largely due to protests that it had a racist intent. Some studies have, nonetheless, pointed to a genetic basis for violence. In 1993, investigators in the Netherlands reported the discovery of what seemed to be an aggression gene, perhaps akin to that which Stephen Mobley claims to run in his family. Biochemical evidence supported the link between the particular mutant, or altered, gene and aggressive behavior.

The gene coded for an enzyme involved in the metabolism of certain chemicals that transmit signals within the brain and nervous system. Urinalysis of subjects in the Dutch study, all of whom were related and demonstrated aggressive and antisocial behavior, showed abnormal levels of metabolic products associated with the enzyme. The affected individuals lacked the ability to produce the enzyme, and this deficiency may have led to their antisocial behavior.

Well-controlled investigations with Drosophila melanogaster, a common fruit fly species, have suggested that a multitude of genes contribute to most behaviors. In a 1992 study, researchers identified Drosophila genes that affect learning, motor behavior, and courtship and mating. Abnormal expression of one gene causes male flies to court other male flies instead of females. Could this finding suggest that genetic factors might determine homosexuality in humans?

Twin studies of homosexual men and women suggest a strong genetic component in homosexual behavior. The experimental design of twin studies is based on the fact that identical twins are genetically matching while fraternal twins are only 50 percent identical. Therefore, identical twins should share inherited traits more often than do fraternal twins. For example, height yields an identical twin correlation of about 0.90 and a fraternal twin correlation of about 0.45, indicating that genetic factors must indeed influence height. Similar experiment designs have yielded significant heritability values for homosexuality.

Work using Restriction Fragment Length Polymorphism (RFLP) markers to correlate the presence of particular gene forms with homosexuality has indicated links between male homosexuality and genetic markers on the X chromosome. Researchers, however, have yet to identify a specific gene for homosexuality.

Other scientists have found evidence that responses to alcohol are heritable in both humans and animals and that multiple genes determine this sensitivity. Some researchers believe a gene that encodes a neurotransmitter receptor protein might contribute to alcoholism. Some studies have found a high frequency of one allele, or variation, of the gene in alcoholic men. It has, however, proved difficult to prove a link to alcoholism.

In addition, scientists have found a gene thought to confer susceptibility to schizophrenia in the genome of a substantial number of those who develop this disorder. Making correlations between the disease and gene frequencies in affected individuals and pinpointing the precise location of the problem genes in the genome still leaves researchers confused.

Scientists are even studying the roles of genes in human tendencies and traits such as depression and intelligence. Because such studies are quite controversial, many people have chosen to dismiss the findings. Almost all the studies of these complex human conditions and qualities have provided only indirect and correlative-not causative-evidence for the roles of specific genes. The evidence suggests that many genes influence such conditions but that linking them to specific genes will prove difficult.

Though we have traditionally phrased the debate about human behavior as "nature versus nurture," it is perhaps more appropriate to call it "nurture given nature" to determine the importance of the environment in the expression of genes without diminishing the important role of genes themselves. Such an encompassing approach has led to novel ways of thinking about the biology of human behavior. Cornell psychology professor Daryl Bem's "exotic becomes erotic," or EBE, theory of the development of human sexual orientation is a good example of consideration of environmental factors in explanations of human behavior.

Bem maintains that adult sexual orientation is actually rooted in childhood preferences for gender-typical or gender-atypical activities and friends. Children who prefer gender-typical activities and friends eventually come to perceive members of the opposite sex as different; this sense of unfamiliarity translates into an exotic view of the opposite sex. Children preferring gender-atypical activities and friends ultimately view members of their own sex as different or exotic. During puberty, feelings of being different evolve into sexual attraction. Thus, what is exotic becomes erotic. Although Bem does not deny the existence of genetic influences on sexual preferences, his theory clearly suggests that heterosexuality and homosexuality are highly determined by an individual's own environment.

Of course, one may create a perpetual argument that it is ultimately an individual's own choice-whether conscious or not-to incorporate the effects of environmental factors into his or her personality or behavior patterns, and that one's propensity to make any such choice is largely founded in genetic makeup. Indeed, the debate about what factors truly contribute to one's behavior will likely continue for a long time to come.

As more research attempts to reveal the precise relationship between genes and behavior, scientists will be presented with a new challenge: to present and discuss their results along with the meanings and implications of such results. The study of behavioral genetics treads on delicate ethical grounds. As has happened in the past, behavioral genetics research data interpreted and usurped by bigoted people easily might be used in the future for a new wave of eugenic thought, leading to extensive discrimination and injustice. To some, reducing behavior to the molecular level may be an offensive oversimplification of human nature-a true blow to personal, religious, and societal concepts of free will and fate. In the end, those who choose to take up the search for intrinsic causes of our behavior will certainly need to evaluate the effects their work will have on society.

BEHAVIOURISM

During the middle of the 20th century', scientific thought again reverted to the mechanical approach and behaviourism reigned throughout America. The behaviourists ignored both genetic effects on behaviour and the ability of animals to engage in flexible problem solving.

The founder of behaviourism, J. B. Watson, stated that differences in the environment can explain all differences in behaviour." He did not believe that genetics had any effect on behaviour. In The Behaviour of Organisms]' the psychologist B. F Skinner wrote that all behaviour could be explained by the principles of stimulus-response and operant conditioning.

The first author visited with Dr. Skinner at Harvard University in 1968. Skinner responded to a question from her about the need for brain research by saying, "We don't need to know about the brain because we have operant conditioning".

Operant conditioning uses food rewards and punishments to train animals and shape their behaviour. In a simple Skinner box experiment, a rat can be trained to push a lever to obtain food when a green light turns on, or to push a lever very quickly to avoid a shock when a red light appears. The signal light is the 'conditioned stimulus." Rats and other animals can be trained to perform a complex sequence of behaviours by chaining together a series of simple operant responses. Skinner believed that even the most complex behaviours can be explained as a series of conditioned responses.

However, a rat's behaviour is very limited in a Skinner box. It's a world with very little variation, and the rat has little opportunity to use its natural behaviours. It simply learns to push a lever to obtain food or prevent a shock. Skinnerian principles explain why a rat behaves a certain way in the sterile confines of a 30 x 30-cm Plexiglas box, but they don't reveal much about the behaviour of a rat in the local dump. Outside of the laboratory, a rat's behaviour is more complex.

INSTINCTS VERSUS LEARNING

Skinner's influence on scientific thinking slowed a bit in 1961 following the publication of "The Misbehaviour of Organisms" by Brelands and Brelands. The paper described how Skinnerian behavioural principles collided with instincts. The Brelands were trained Skinnerian behaviourists who attempted to apply the strict principles of operant conditioning to animals trained at fairs and carnivals.

Ten years before this classic paper, the Brelands wrote, we are wholly affirmative and optimistic that principles derived from the laboratory can be applied to the extensive control of animal behaviour under non laboratory condition. However, by 1961, after training more than 6000 animals as diverse as reindeer, cockatoos, raccoons, porpoises, and whales for exhibition in zoos, natural history museums, department store displays, fair and trade convention exhibits, and television, the Brelands wrote a second article featured in the American Psychologist, which stated, "our backgrounds in behaviourism had not prepared us for the shock of some of our failures".

One of the failures occurred when the Brelands tried to teach chickens to stand quietly on a platform for 10 to 12 seconds before they received a food reward. The chickens would stand quietly on a platform in the beginning of training; however, once they learned to associate the platform with a food reward, half (50%) started scratching the platform, and another 25% developed other behaviours, such as pecking the platform.

The Brelands salvaged this disaster by developing a wholly unplanned exhibit involving a chicken that turned on a juke box and danced. They first trained the chickens to pull a rubber loop which turned on some music. When the music started, the chickens would jump on the platform and start scratching and pecking until the food reward was delivered.

This exhibit made use of the chicken's instinctive food-getting behaviour. The first author remembers as a teenager seeing a similar exhibit, at the Arizona State Fair, of a piano-playing chicken in a little red barn. The hen would peck the keys of a toy piano when a quarter was put in the slot and would stop when the food came down the chute. This exhibit also worked because it was similar to a Skinner box in the laboratory.

The Brelands experienced another classic failure when they tried to teach raccoons to put coins in a piggy bank. Because raccoons are adept at manipulating objects with their hands, this task was initially easy. As training progressed, however, the raccoons began to rub the coins before depositing them in the bank. This behaviour was similar to the washing behaviour raccoons do as instinctive food-getting behaviour. The raccoons at first had difficulty letting go of the coin and would hold and rub it. However, when the Brelands introduced a second coin, the raccoons became almost impossible to train. Rubbing the coins together 'in a most miserly fashion]' the raccoons got worse and worse as time went on. The Brelands concluded that the innate behaviours were suppressed during the early stages of training and sometimes long into the training, but as training progressed, instinctive food-getting behaviours gradually replaced the conditioned behaviour. The animals were unable to override their instincts and thus a conflict between conditioned and instinctive behaviours occurred.

ETHOLOGY

While Skinner and his fellow Americans were refining the principles of operant conditioning on thousands of rats and mice, ethology was being developed in Europe. Ethology is the study of animal behaviour in natural environments and the primary concern of the ethologists is instinctive or innate behaviour.

Essentially, ethologists believe that the secrets to behaviour are found in the animals genes and in the way the genes have been modified during evolution to deal with particular environments. The ethological trend originated with Whitman, who regarded instincts as congenital reactions which are so constant and characteristic for each species that, like morphological structures, they may he of taxonomic significance.

A similar opinion was held by Heinroth. He trained newly hatched fledglings in isolation from adult birds of their own species and found that instinctive movements such as preening, shaking, and scratching were performed by young birds without observing other birds.

Understanding the mechanisms and programming that result in innate behavioural patterns and the motivations behind why animals behave the way they do is the primary focus of ethologists. Konrad Lorenz and Niko Tinbergen cataloged the behaviour of many animals in their natural environments. Together they developed the ethogram. An ethogram is a complete listing of all the behaviours that an animal performs in its natural environment. The ethogram includes both innate and learned behaviours.

An interesting contribution to ethology came from studies on egg-rolling behaviour in the greylag goose. He observed that when a brooding goose notices an egg outside her nest, an innate instinctive programme is triggered to retrieve it. The goose fixates on the egg, rises to extend her neck and bill out over it, then gently rolls it back to the nest. This behaviour is performed in a highly mechanical way If the egg is removed as the goose begins to extend her neck, she still completes the pattern of rolling the nonexistent egg back to the nest.

Lorenz and Tinbergen termed this a 'fixed action pattern." Remarkably, Tinbergen also discovered that brooding geese can be stimulated to perform egg rolling on such items as beer cans and baseballs. The fixed action pattern of rolling the egg back to the nest can be triggered by anything outside the nest that even marginally resembles an egg.

Tinbergen realised that geese possess a genetic-releasing mechanism for this fixed action pattern. Lorenz and Tinbergen called the object that triggers the release of a fixed action pattern "sign stimuli." When a mother bird sees the gaping mouth of her young, it triggers the maternal feeding behaviour and the mother feeds her young. The gaping mouth is another example of sign stimuli that acts as a switch and turns on the genetically determined programme.

Ethologists also explained the innate escape response of newly hatched goslings. When goslings are tested with a cardboard silhouette in the shape of a hawk moving overhead, it triggers a characteristic escape response. The goslings will crouch or run. However, when the silhouette is reversed to look like a goose, there is no effect. Several members of the research community doubted the existence of such a hard-wired instinct because other scientists failed to repeat these experiments.

Recently Canty and Gould repeated the classic experiments and explained why the other experiments failed. In the first place, goslings only respond to the silhouette when they are under 7 days old. Second, a large silhouette which casts a shadow must be used; third, goslings respond to the perceived predator differently depending on the circumstances. For example, birds tested alone try to run away from the hawk silhouette and birds reared and tested in groups tend to crouch.

Nevertheless, fear is likely to be the basis of the response. Ducklings were shown to have higher heart rate variability when they saw the hawk silhouette. Research by Balaban indicates that species-specific vocalisations and head movements in chickens and quail are controlled by distinct cell groups in the brain. To prove this, Balaban transplanted neural tube cells from developing quail embryos into chicken embryos. Chickens hatched from the transplanted eggs exhibited species-specific quail songs and bobbing head movements.

Do similar fixed action patterns occur in mammals? Fentress conducted an experiment on mice which clearly showed that animals have instinctive species-specific behaviour patterns which do not require learning. Day-old baby mice were anesthetised and had a portion of their front legs amputated. Enough of the leg remained that the mice could easily walk.

The operations were performed before the baby mice had fully coordinated movements so there was no opportunity for learning. When the mice became adults, they still performed the species-specific face-washing behaviour; normal mice close their eye just prior to the foreleg passing over the face, and in the amputees the eye still closed before the nonexistent paw hit it. The amputees performed the face- washing routine as if they still had their paws.

SCIENCE OF BEHAVIOUR TODAY

Two years after the Brelands article, Jerry Hirsh at the University of Illinois wrote a paper emphasising the importance of studying individual differences. He wrote, "Individual differences are no accident. They are generated by properties of organisms as fundamental to behaviour science as thermodynamic properties are to physical science."

Today, scientists recognise the contributions of both the Skinnerian and the ethologists approach to understanding behaviour Modern neuroscience supports Darwin's view on behaviour. Bird and mammal brains are constructed with the same basic design. They all have a brain stem, limbic system, cerebellum, and cerebral cortex.

The cerebral cortex is the part of the brain used for thinking and flexible problem solving. The major difference between the brains of people and animals is in the size and complexity of the cortex. Primates have a larger and more complex cortex than a dog or a pig; pigs have a more complex cortex than a rat or a mouse.

Furthermore, all animals possess innate species-specific motor patterns which interact with experience and learning in the formation of behaviour. Certain behaviours in both wild and domestic animals are governed largely by innate (hard-wired) programmes; however, experiencing and learning are the most important factors in other behaviours.

A basic principle to remember is that animals with large, complex brains are less governed by innate behaviour patterns. For example, bird behaviour is governed more by instinct than that of a dog, whereas an insect would have more hard-wired behaviour patterns than that of a bird. This principle was clear to Yerkes who wrote:

> Certain animals are markedly plastic or voluntary in their behaviour, others are as markedly fixed or instinctive. In the primates plasticity has reached its highest known stage of development; in the insects fixity has triumphed, instinctive action is predominant.

The ant has apparently sacrificed adapt-ability to the development of ability to react quickly, accurately and uniformly in a certain way Roughly, animals might he separated into two classes: those which are in high degree capable of immediate adaptation to their conditions, and those that are apparently automatic since they depend upon instinct tendencies to action instead of upon rapid adaptation.

GENETICS AND EXPERIENCE

Some behaviour patterns are similar between different species, and some are found only in a particular species. For example, the neural programmes that enable animals to walk are similar in most mammals. On the other hand, courtship rituals in birds are very species-specific. Some innate behaviour patterns are very rigid and experience has little effect on them; other instinctive behaviours can be modified by learning and experience.

The flehmann, or lip curl response of a bull when he smells a cow in estrus, and the kneel-down (lordois) posture of a rat in estrus are examples of behaviours that are rigid. Suckling of the mother by newborn mammals is another example of a hard-wired behavioural system. Suckling behaviour does not vary Newborn mammals suckle almost anything put in their mouth.

An example of an innate behaviour that is affected by learning is burrowing behaviour in rats. Boice found that wild Norway rats and albino laboratory rats both dig elaborate burrows. Learning has some effect on the efficiency of burrowing, but the configuration of the burrows was the same for both the wild and domestic rats. The albino laboratory rats dug excellent burrows the first time they were exposed to an outdoor pen.

Nest building in sows is another example of the interaction between instinct and learning. When a sow is having her first litter, she has an uncontrollable urge to build a nest. Nest building is hard-wired and hormonally driven because prostaglandin F2a injections will induce it in sows. However, sows earn from experience how to build a better nest with each successful litter.

Other behaviours are almost entirely learned. Some seagulls learn to drop shellfish on rocks to break them open, while others drop them on the road and let cars break them open. Many animals ranging from apes to birds use tools to obtain food. Griffin and Dawkins provide many examples of complex learned behaviours and flexible problem solving in animals.

Innate behaviours used for finding food, such as grazing, scavenging, or hunting, are more dependent on learning than behaviours used to consume food. Mating behaviour, nesting, eating, and prey-killing behaviours tend to be governed more by instinct. The greater dependence on learning to find food makes animals in the wild more flexible and able to adapt to a variety of environments.

Behaviours used to kill or consume food can be the same in any environment. Mayr called these different behavioural systems "open" or "closed" to the effects of experience. A lion hunting her prey is an example of an open system. The hunting female lion recognizes her prey from a distance and carefully stalks her approach. Herrick wrote, "the details of the hunt vary every time she hunts; therefore, no combination of simple reflex arcs laid down in the nervous system will be adequate to meet the infinite variations of the requirements for obtaining food.'

Complex Interactions

Some of the interactions between genetics and experience have very complex effects on behaviour. In birds, the chaffinch learns to sing its species-specific song even when reared in a sound-proof box where it is unable to hear other birds.

However, when chaffinches are allowed to hear other birds sing, they develop a more complex song. The basic pattern of canary song emerges even in the absence of conspecific (flock-mate) auditory model. Young canaries imitate the song of adult canaries they can hear, and when reared in groups they develop song patterns that they all share. Many birds, such as the white crowned sparrow chaffinch, and parrot, can develop local song dialects.

Sparrows are able to learn songs by listening to recordings of songs with either pure tones or harmonic overtones. Birds trained with harmonic overtones learned to sing songs with harmonic overtones, but 1 year later, 85% of their songs reverted back to innate pure tone patterns.

Further experiments by Mundinger attempted to determine the relative contribution of genetics and learning in bird song. Inbred lines of roller and border canaries were used in this study along with a hybrid cross of the two. The rollers were cross fostered to border hens and vice versa to control for effects of maternal behaviour. The roller and border males preferred to sing innate song patterns instead of copying their tutors. The hybrids preferred to learn some of both songs. Furthermore, canaries are capable of learning parts of an alien song but have a definite preference for their own songs.

Comparing these animals to those in Brelands and Brelands exhibits, birds can be trained to sing a different song, but genetically determined patterns have a strong tendency to override learning. In reviewing all this literature, it became clear that innate patterns in mammals can be overridden. Unfortunately the animals tend to revert back to innate behaviour patterns.

PARADOX OF NOVELTY

Novelty is anything new or strange in an animal's environment. Novelty is a paradox because it is both fear-provoking and attractive. Paradoxically it is most fear-provoking and attractive to animals with a nervous, excitable temperament.

Skinner wrote that a flighty animal such as the pronghorn antelope will approach a person lying on the ground waving a red flag. Einarsen further observed that some wild animals will approach various large, dangerous objects such as a power steam shovel. In more recent studies, Kruuk also observed attraction and reaction to novelty in Thompson's gazelles in Africa. In small groups, Thompson's gazelles are most watchful for predators.

Animals that survive in the wild by flight are more attentive to novelty than more placid animals. Gazelles can also distinguish between a dangerous hunting predator and one that is not hunting. The most dangerous predators attract the highest degrees of attraction in the Thompson's gazelle. They often move close to a cheetah when the cheetah is not hunting. Furthermore, when predators walk through a herd of Thompson's gazelles, the size of the flight zone varies depending on the species of predator.

Reaction to Novelty

Confronted with sudden novelty, highly reactive animals are more likely to have a major fear reaction. Examples of sudden novelty include being placed in a new cage, transport in a strange vehicle, an unexpected loud noise, or being placed in an open field.

Using various experimental environments, Hennessy and Levine found that rats show varying degrees of stress and stress hormone levels proportional to the degree of novelty of the environment they are placed in; a glass jar is totally novel in appearance compared to the lab cage to which the animal was accustomed. Being placed in a glass jar was more stressful for rats than a clean lab cage with no bedding.

Livestock and Poultry Reaction to Novelty

Studies of the reaction to novelty in farm animals have been conducted by Moberg and Wood, Stephens and Toner and Dantzer and Mormede. When calves are placed in an open field test arena that is very dissimilar from their home pen, they show the highest degrees of stress.

Calves raised indoors were more stressed by an outdoor arena and calves raised outdoors were more stressed by an indoor arena. The second author is painfully familiar with similar responses in horses. When horses are taken to the mountains for the first time, a well-trained riding horse that is accustomed to many different show rings may panic when it sees a butterfly or hears a twig snapping on a mountain trail.

Genetic Factors and the Need for Novelty

In mammals and birds, normal development of the brain and sense organs requires novelty and varied sensory input. Nobel prize winning research of Hubel and Wiesel showed that the visual system is permanently damaged if kittens do not receive varied visual input during development. When dogs are raised in barren and nonstimulating environments they are also more excitable. Schultz stated, "when stimulus variation is restricted central regulation of threshold sensitivities will function to lower sensory thresholds."

Krushinski studied the influence of isolated conditions of rearing on the development of passive defense reactions (fearful aggression) in dogs and found that the expression of a well-marked fear reaction depends on the genotype of the animal. Airedales and German shepherds were reared under conditions of freedom (in homes) and in isolation (in kennels).

Krushinski found that the passive defense reaction developed more acutely and reached a greater degree in the German shepherds kept in isolation compared to the Airedales. In general, animals reared in isolation become more sensitive to sensory stimulation because the nervous system attempts to readjust for the previous lack of stimulation.

In an experiment with chickens, Murphy found that chicks from a flighty genetic line were more likely to panic when a novel ball was placed in their pen, but they were also more attracted to a novel food than birds from a calm line. Cooper and Zubeck, and Henderson found that rats bred to be dull greatly improved in maze learning when housed in a cage with many different objects; however, enriched environments had little effect on the rats bred for high intelligence.

Greenough and Juraska found that rearing rats in an environment with many novel objects improves learning and resulted in increased growth of dendrites, which are nerve endings in the brain.

Pigs raised in barren concrete pens also seek stimulation. Piglets allowed to choose between a familiar object and a novel object prefer the novel object. Pigs raised on concrete are strongly attracted to novel objects to chew on and manipulate. The first author has observed that nervous, excitable hybrid pigs often chew and bit vigorously on boots or coveralls. This behaviour is less common in placid genetic lines of pigs.

Although hybrid pigs are highly attracted to novelty, tossing a novel object into their pen will initially cause a strong flight response. Compared to calm genetic lines, nervous-hybrid pigs pile up and squeal more when startled. Pork producers report that nervous, fast-growing, lean hybrid pigs also tail-bite other pigs more often than calmer genetic lines Of pigs. jail biting occurs more often when pigs are housed on a concrete slatted floor which provides no opportunity for rooting.

Strong attraction or strong reaction to novelty has also been observed by the first author in cattle. Cattle will approach and lick a piece of paper laying on the ground when they can approach it voluntarily. However, the same piece of paper blowing in the wind may trigger a massive flight response.

Practical experience by both authors suggests that highly reactive horses are more likely to engage in vices such as cribbing or stall weaving when housed in stalls or runs where they receive little exercise. Denied variety and novelty in their environments, highly reactive animals adapt poorly compared to animals from calmer genetic lines. In summary, in both wild and domestic animals novelty is both highly feared and necessary Novelty is most desirable when animals can approach it slowly. Unfortunately, novelty is also fear-provoking when animals are suddenly confronted with it.

TEMPERAMENT

In animals as diverse as rats, chickens, cattle, pigs, and humans, genetic factors influence differences in temperament. Some individuals are wary and fearful and others are calm and placid. Boissy stated, fearfulness is a basic psychological characteristic of the individual that predisposes it to perceive and react in a similar manner to a wide range of potentially frightening events. In all animals, genetic factors influence reactions to situations which cause fear; therefore, temperament is partially determined by an individual animal's fear response.

Rogan and LeDoux suggest that fear is the product of a neural system that evolved to detect danger and that it causes an animal to make a response to protect itself. Plomin and Daniels found a substantial genetic influence on shyness (fearfulness) in human children. Shy behaviour in novel situations is considered a stable psychological characteristic of certain individuals. Shyness is also suggested to be among the most heritable dimensions of human temperament throughout the life span.

In an experiment designed to control for maternal effects on temperament and emotionality, Broadhurstconducted cross-fostering experiments on Maudsley Reactive (MR) and Non Reactive (MNR) rats. These lines of rats are genetically selected for high or low levels of emotional reactivity

The results showed that maternal effects were not great enough to completely mask the temperament differences between the two lines. Maternal effects can affect temperament, but they are not great enough to completely change the temperament of a cross-fostered animal which has a temperament that is very different from that of the foster mother.

Measuring Fear-Based Behaviours

One method of testing fearfulness is the open field test. Sudden placement of an animal in an open field test arena is used to measure differences in fearfulness. Open field testing has shown differences in fearfulness between different genetic lines of animals. The test arena floor is usually marked in a grid to measure how much the animals walk around and explore.

Huck and Price showed that domestic rats are less fearful and will walk round the open held more than wild rats. Price and Loomis explained that some genetic strains of rats are less fearful and explore an open field arena more than others. Eysenck and Broadhurst found that rodents with high emotional reactivity are more fearful and explore the open field less compared to placid genetic lines.

In their study of genetic effects on behaviour, Fuller and Thompson found that "simply providing the same defined controlled environment for each genetic group is not enough. Conditions must not only be uniform for all groups, but also favorable to the development of the behaviour of interest."

For example, in wartime Russia, Krushinski investigated the ability of dogs to be trained for the antitank service or as trail dogs trained to track human scent. The dogs were tied to a spike driven into the ground, and the person who regularly looked after them would let them lick from a bowl of food and then summon the dog to follow the man as he retreated 10 to 15 meters. The dog's activity was measured with a pedometer for the next 2 minutes.

The most active dogs were found to be the best antitank dogs. They were also fearless. In the antitank service, dogs were trained to run up to a tank and either run along side of it or penetrate under the caterpillars of the tank. In order to do this, the dogs had to overcome their natural fear of a tank moving toward them at high speed.

The less active dogs (as measured by the pedometer) were found to make the best trailer dogs. They slowly followed a trail and kept their noses carefully to the scent while negotiating the corners and turns on the trail. The more active dogs trailed at too high a speed and often jumped the corners and turns in the trail, which sometimes resulted in switching to another trail.

Mahut demonstrated an example of differences in fear responses between beagles and terriers. When frightened, beagles freeze and terriers run around frantically In domestic livestock, measuring fear reactions during restraint or in an open field test has revealed differences in temperament both between breeds and between individuals within a breed.

The fearful, flighty animals become more agitated and struggle more violently when restrained for vaccinations and other procedures. Fear is likely to be the main cause of agitation during restraint in cattle, horses, pigs, and chickens.

Species Differences in Fear Reactions

In an open field test, frightened rodents often stay close to the arena walls, whereas frightened cattle may run around wildly and attempt to escape. Rodents stay close to the walls because they naturally fear open spaces, whereas cattle run around wildly because they fear separation from the herd. This is an example of differences between species in their response to a similar fear- provoking situation. Fear can be manifested in many different ways. For example, a frightened animal may run around frantically and try to escape in one situation, while in another situation the same animal may freeze or limit its movement. Chickens often freeze when handled by humans. Jones called this "tonic immobility." The chickens become so frightened that they cannot move. Forceful capture of wild animals can sometimes inflict fatal heart damage. Wildlife biologists call this capture myopathy In summary, much is known about the complex phenomenon of fear, but many questions still remain.

BIOLOGICAL BASIS OF FEAR

Genetics influences the intensity of fear reactions. Genetic factors can also greatly reduce or increase fear reaction in domestic animals. Research in humans has clearly revealed some of the genetic mechanisms which govern the inheritance of anxiety.

LeDoux and Rogan and LeDoux state that all vertebrates can be fear-conditioned. Davis recently reviewed studies on the biological basis of fear. Overwhelming evidence points to the amygdala as the fear center in the brain. A small bilateral structure located in the limbic system, the amygdala is where the triggers for flight or fight" are located. Electrical stimulation of the amygdala is known to increase stress hormones in rats and cats; destroying the amygdala can make a wild rat tame and reduce its emotionality.

Destroying the amygdala also makes it impossible to provoke a fear response in animals. Blanchard and Blanchard showed that rats lose all of their fear of cats when the amygdala is lesioned. Furthermore, when a rat learns that a signal light means an impending electric shock, a normal response is to freeze. Destroying the amygdala will eliminate this response. Finally, electrical stimulation of the amygdala makes humans feel fearful. Animal studies also show that stimulation of the amygdala triggers a pattern of responses from the autonomic nervous system similar to that found in humans when they feel fear.

Heart rate, blood pressure, and respiration also change in animals when the flight or fight response is activated. All these autonomic functions have neural circuits to the amygdala. Fear can be measured in animals by recording changes in autonomic activity In humans, Manuck and Schaefer found tremendous differences in cardiovascular reactivity in response to stress, reflecting a stable genetic characteristic of individuals.

Fearfulness and Instinct

Fearfulness and instinct can conflict. This principle was observed firsthand by the second author during his experience raising Queensland Blue Heeler dogs. Annie's first litter was a completely novel experience because she had never observed another dog giving birth or nursing pups. She was clearly frightened when the first pup was born and it was obvious that she did not know what the pup was; however, as soon as she smelled it her maternal instinct took over and a constant uncontrollable licking began.

Two years later, Annie's daughter Kay had her first litter. Kay was more fearful than her mother and her highly nervous temperament overrode her innate licking programme. When each pup was born Kay ran wildly around the room and would not go near them. The second author had to intervene and place the pups under Kay's nose; otherwise, they may have died. Kay's nervous temperament and fearfulness were a stronger motivation than her motherly instinct.

NERVOUS SYSTEM AND ENVIRONMENT

Raising young animals in barren environments devoid of variety and sensory stimulation will have an effect on development of the nervous system. It can cause an animal to be more reactive and excitable when it becomes an adult. This is a long-lasting, environmentally induced change in how the nervous system reacts to various stimuli. Effects of deprivation during early development are also relatively permanent.

Melzak and Burns found that puppies raised in barren kennels developed into hyperexcitable adults. In one experiment the deprived dogs reacted with "diffuse excitement" and ran around a room more than control dogs raised in homes by people.

Presenting novel objects to the deprived dogs also resulted in diffuse excitement." Furthermore, the EEGs of the kennel-raised dogs remained abnormal even after they were removed from the kennel. Simons and Land showed that the somatosensory cortex in the brains of baby rats will not develop normally if sensory input was eliminated by trimming their whiskers. A lack of sensory input made the brain hypersensitive to stimulation. The effects persisted even after the whiskers had grown back.

Development of emotional reactivity of the nervous system begins during early gestation. Denenberg and Whimbey showed that handling a pregnant rat can cause her offspring to be more emotional and explore less in an Open field compared to control animals. This experiment is significant because it shows that handling the pregnant mother had the opposite effect on the behaviour of the infant pups.

Handling and possibly stressing the pregnant mothers changed the hormonal environment of the fetus which resulted in nervous offspring. However, handling newborn rats by briefly picking them up and setting them in a container reduced emotional reactivity when the rats became adults. The handled rats developed a calmer temperament.

The adrenal glands are known to have an effect on behaviour. The inner portions of the adrenals secrete the hormones adrenaline and noradrenaline, while the outer cortex secretes the gender hormones androgens and oestrogens (reproductive hormones), and various corticosteroids (stress hormones).

Yeakel and Rhoades found that Hall's emotional rats had larger adrenals and thyroids compared to the nonemotional rats. Richter found a decrease in the size of the adrenal glands in Norway rats accompanied by domestication. Several line and strain differences have been found since these early reports. Furthermore, Levine and Levine *et al.* showed that brief handling of baby rats reduces the response of the adrenal gland to stress.

Changing Reactivity versus Taming

Adult wild rats can be tamed and become accustomed to handling by people. This is strictly learned behaviour. Taming full-grown wild animals to become accustomed to handling by people will not diminish their response to a sudden novel stimulus. This principle was demonstrated by Grandin *et al.* in training wild antelope at the Denver Zoo for low-stress blood testing.

Nyala are African antelope with a hair-trigger flight response used to escape from predators. During handling in zoos for veterinary treatments, nyala are often highly stressed and sometimes panic and injure themselves. Over a period of 3 months, Grandin *et al.* trained nyala to enter a box and stand quietly for blood tests while being fed treats. Each new step in the training had to be done slowly and carefully Ten days were required to habituate the nyala to the sound of the doors on the box being closed.

All the training and petting by zoo keepers did not change the nyala's response to a sudden, novel stimulus. When the nyala saw repairman on the barn roof they suddenly reacted with a powerful fear response and crashed into a fence. They had become accustomed to seeing people standing at the perimeter of the exhibit, but people on the roof was novel and very frightening. Sudden movements such as raising a camera up for a picture also caused the nyala to flee.

Domestic versus Wild

Wild herding species show much stronger fear responses to sudden novelty compared to domestic ruminants such as cattle and sheep. Domestic ruminants have attenuated flight responses due to years of selective breeding. Wild ruminants will learn to adapt in captivity and associate people with food, but when frightened by some novel stimulus they are more likely to panic and injure themselves. This is especially likely if they are prevented from fleeing by a fence or other barrier.

Principles for training and handling all herding animals are basically similar. Training procedures used on flighty antelope or placid domestic sheep are the same. The only difference is the amount of time required. Grandin demonstrated this by training placid Suffolk sheep to voluntarily enter a tilting restraining device in one afternoon, but it took 3 months to train the nyala. In summary, experience can affect behaviour in two basic ways: by conventional learning or by changing nervous system reactivity Most importantly, environmental conditions have the greatest effect on the nervous systems of young animals.

NEOTENY

Neoteny is the retention of the juvenile features in an adult animal. Genetic factors influence the degree of neoteny in individuals. Neoteny is manifested both behaviourally and physically In the forward to "The Wild Canids", Conrad Lorenz adds a few of his observations on neoteny and the problems of domestication:

> The problems of domestication have been an obsession with me for many years. On the one hand I am convinced that man owes the life-long persistence of his constitutive curiosity and explorative playfulness to a partial neoteny which is indubitably a consequence of domestication In a curiously analogous manner does the domestic dog owe its permanent attachment to its master to a behavioural neoteny that prevents it from ever wanting to be a pack leader On the other hand, domestication is apt to cause an equally alarming disintegration of valuable behavioural traits and an equally alarming exaggeration of less desirable ones.

Infantile characteristics in domestic animals are discussed by Price, Lambooij and van Putten, Coppinger and Coppinger, Coppinger and Scheider, and Coppinger *et al.* The shortened muzzle in dogs and pigs is an example. Domestic animals have been selected for a juvenile head shape, shortened muzzles, and other features. Furthermore, retaining juvenile traits makes animals more tractable and easy to handle. The physical changes are also related to changes in behaviour.

Genetic studies point to the wolf as the ancestor of domestic dogs. During domestication, domestic dogs have retained many of the infant wolf behaviours. For example, wolf pups bark and yap a lot but adult wolves rarely bark; domestic dogs bark a lot. Wolves have hard-wired instinctive behaviour patterns that determine dominance or submission in social relationships. In domestic dogs, the ancestral social behaviour patterns of the wolf are fragmented and incomplete.

Frank and Frank observed that the rigid social behaviour of the wolf has disintegrated into "an assortment of independent behavioural fragments." Malamutes raised with wolf pups fail to read the social behaviour signals of the wolf pups. Further comparisons found that the physical development of motor skills is slower in the malamute.

Goodwin *et al.* studied 10 different dog breeds which ranged from German shepherds and Siberian huskies to bulldogs, cocker spaniels, and terriers. They found that the breeds which retained the greatest repertoire of wolf-like social behaviours were the breeds that physically resembled wolves, such as German shepherds and huskies. Barnett *et al.* and

Price both conclude that experience can also cause an animal to retain juvenile traits. Gould also considered the effects of neoteny and stated that neoteny is determined by changes in a few genes that determine the timing of different developmental stages.

OVERSELECTION FOR SPECIFIC TRAITS

Countless examples of serious problems caused by continuous selection for a single trait can be found in the medical literature. People with experience in animal husbandry know that overselection for single traits can ruin animals. Good dog breeders know this. Sometimes traits that appear to be unrelated are in fact linked. Wright demonstrated this clearly by continuous selection for hair colour and hair patterns in inbred strains of guinea pigs. Depressed reproduction resulted in all the strains. Furthermore, differences in temperament, body conformation, and the size and shape of internal organs were found.

Continuous selection for a calm temperament in foxes resulted in negative effects on maternal behaviour and neurological problems. The fox experiments also found graded changes in many traits over several years of continuous selection for tame behaviour. Physiological and behavioural problems increased with each successive generation. In fact, some of the tamest foxes developed abnormal maternal behaviour and cannibalised their pups. Belyaev *et al.* called this "destabilising selection," in contrast to "stabilising selection" found in nature.

There are also countless examples in the veterinary medical literature of abnormal bone structure and other physiological defects caused by overselecting for appearance traits in dog breeds. The abnormalities range from bulldogs with breathing problems to German shepherds with hip problems.

Overselection in Livestock

Single-minded selection for production traits such as rapid gain and leanness resulted in pigs and cattle with more excitable temperaments. Compared to the older genetic lines with more hack fat, observations by the first author on thousands of pigs indicate that lean hybrids are more excitable and difficult to drive through races. Lean hybrid pigs also have a greater startle response. Separating a single animal from the group is more difficult.

Links between Different Traits

Casual observations by the first author also indicate that the most excitable, flighty pigs and cattle have a long, slender body with fine bones. Some of the lean hybrid pigs have weak legs and a few of the normally brown-eyed pigs now have blue eyes. Blue eyes are often associated with neurological problems. Furthermore, pigs and cattle with large, bulging muscles often have a calmer temperament than lean animals with less muscle definition. However, animals with the muscle hypertrophy trait (double muscling) have a more excitable temperament. Double muscling is extreme abnormal muscling and it might have the opposite effect on temperament compared to normal muscling.

Another example of apparently unrelated traits being linked is deafness in dogs of the pointer breed selected for nervousness. There appears to be a relationship between thermoregulation and aggressiveness. Wild mice selected for aggressiveness used larger amounts of cotton to build their nests than mice selected for low aggression. This effect occurred in both laboratory and wild Strains of mice.

Researchers using high-tech "knockout" gene procedures have been frustrated by the complexity of genetic interactions. In this procedure, genes are knocked out in a gene-targeting procedure whereby a gene is prevented from performing its normal function. The knockout experiments have shown that blocking different genes can have unexpected effects on behaviour. In one experiment, superaggressive mice were created when genes involved with learning were inactivated. The mutant mice had little or no fear and fought until they broke their backs. In another experiment the knockout mutants demonstrated normal behaviour until they had pups, and failed to care for them.

In still another experiment, Konig *et al.* disabled the gene that produces enkephalin (a brain opioid substance) and found unexpected results. Enkephalin is a substance normally involved in pain perception; however, the mice that were deficient in this substance were very nervous and anxious. They ran frantically around their cages in response to noise. The bottom line conclusion from several different knockout experiments is that changing one gene has unexpected effects on other systems. Traits are linked, and it may be impossible to completely isolate single gene effects.

Researchers warn that one must be careful not to jump to a conclusion that they have found an "aggression gene" or a "maternal gene" or an "anxiety gene." To use an engineering analogy, one would not conclude that they had found the "picture center" in a television set after they cut one circuit inside the set that ruined the picture.

Random Factors

Behavioural geneticists have discovered that it is impossible to completely control variation in some traits. Gartner found that breeding genetically similar inbred lines of rats failed to stop weight fluctuations. Even under highly standardised laboratory conditions, body weights continued to fiuctuate between animals. Pig breeders have also observed that commercially bred hybrid lines of pigs do not gain weight at the same rate. Random unknown factors affect variability even in genetically identical animals. Factors in utero may be one cause; the other causes are unknown. Darrel Tatum and his students at Colorado State University found both body conformation and meat quality variation in cattle which were 50% English (Bos taurus) and 50% Brabman (Bos indicus). Some animals had more Brahman characteristics, with larger humps and longer ears than others; the body conformation of many of the animals was not half English and half Brahmatn. The characteristics of the meat varied as well; animals that looked more Brahman had tougher meat. The animals had about 10% variation from the body shape and meat characteristics of Brahman half-bloods.

Gartner concluded that up to 90% of the cause of random variability cannot be explained by differences in the animals' physical environment. In both mice and cattle, random factors affected body weights. Gartner believes that the random factors may have their influence either before or shortly after fertilisation. The interactions between environmental and genetic factors are complex. Both an animals' genetic makeup and its environment determine how it will behave.

BEHAVIOURAL GENETICS : FUTURE PROSPECTS

During its first 30 years, from roughly 1960 to 1990, the modern discipline of behavioural genetics was based almost entirely on twin and family studies. Those studies made a strong case for the importance of genes in behaviour, but the connection always remained loose and statistical. Only in rare cases could a direct connection between a particular gene or set of genes and a particular behaviour be made.

In the past decade and a half, all that has changed with the introduction of bioinformatics, genetic engineering and other techniques that allow researchers to measure, analyse and manipulate genetic material rapidly and easily. These techniques have changed the composition of the field of behavioural genetics, engaging the interest of new groups of researchers beyond psychology—molecular biologists, medical doctors and others—who had previously seen behaviour as too slippery for biological research.

This shift took place during a time when interest in genetics was exploding. The announcement in 2000 of a completed draft of the human genome—the total complement of genes found in the nucleus of each human cell—and the 50th anniversary of Francis Crick and James Watson's discovery of the structure of DNA in 2003 marked the high points.

Today, expectations of quick rewards from the use of these new techniques are lower than they were during the first flush of excitement. It is now clear that a single gene for complex disorders such as depression is unlikely to exist, let alone be found, even with the most sophisticated methods. Complex behavioural traits, researchers are finding, are influenced by tens if not hundreds of genes, each interacting with the environment and each other in unpredictable ways.

Nonetheless, behavioural genetics continues to hold out the promise of better understanding the biological basis of behaviour—hence the field receives strong support from the National Institutes of Health and other grant-making institutions concerned with the intersection of behaviour and health.

The new techniques have not replaced the classic methods in behavioural genetics: twin and family studies that used genetic relatedness to search for genes associated with behaviour. In fact, twin studies remain one of the best ways of identifying genetic markers linked to complex behavioural traits, according to researchers such as John DeFries, PhD, founder of the journal Behavioural Genetics and former director of the Institute for Behavioural Genetics. Increasingly, however, such studies are being used not as end-points in themselves, but as stepping stones toward molecular genetics studies that can identify particular genes and their functions.

Ten years ago, before the Human Genome Project and the proliferation of inexpensive genetic tests, a researcher studying a particular behavioural disorder might have had access to tests for three or four gene. Now you click on the Internet and you can find information for the whole genome. Such information is now available not just for the human genome, but also for common laboratory animals such as mice. This flood of data means that the ability to gather, organise and analyse biological information is becoming increasingly critical. Labs recently hires bioinformatics specialist to stay up-to-date on methods for mining the gigabytes of data now available. New techniques are also providing scientists with ways of directly manipulating genes in animals and observing the altered genes' effects on behaviour. Mice have proved to be especially amenable to such manipulation. There are now thousands of different strains of single-gene mutants and "knockout" mice—animals in which a single gene has been altered or disabled.

APA genetics task force member Jeanne Wehner, PhD, of the University of Colorado at Boulder, is among those who have studied such knockouts. Although her training is in biochemistry, she and her laboratory do work that is primarily psychological. Using tests of learning and cognition, they look for behavioral differences in strains of genetically manipulated mice. One knockout-mouse strain they have studied is missing the gene for protein kinase C gamma, a cellular "second-messenger" that communicates between surface receptors and the internal machinery of neurons in the brain and spinal cord.

Like standard mice, these knockouts can be trained to respond to a stimulus in exchange for a reward. However, in experiments where rewards are given for withholding a response—rather than for responding immediately after each stimulus—the mice tend to perform poorly. This, together with their tendency to drink more alcohol than standard mice, is taken as indication of their impulsivity.

Wehner and her colleagues at the University of Colorado are now trying to flesh out the links between protein kinase C gamma and its possible effects on human behaviours such as drug abuse and alcoholism. One set of studies, led by a member of Wehner's lab, neuroscientist Barbara Bowers, PhD, is examining the effects of protein kinase C within the cell. She is testing the hypothesis that the missing gene affects serotonin receptors, which are known to be involved in emotion and motivation.

Another set of studies, led by Marissa Ehringer, PhD, a human genetics researcher also at the Institute for Behavioural Genetics, is trying to bridge the gap between animals and humans. As part of a larger project on adolescent anti-social behaviour, Ehringer is looking for evidence that humans show variation in the gene for protein kinase C gamma and whether that variation has consequences for behaviour.

As with much of today's behavioural genetics research, the protein kinase C studies would be impossible without the collaboration of people from a variety of disciplines: the biologists who created the knockout animals, the neuroscientists and psychologists who designed and implemented the animal behaviour studies, and the psychologists and medical geneticists who are looking for genetic variation in humans.

The proliferation of new techniques has raised expectations of what behavioural genetics can do. But, as many researchers are quick to note, those expectations can sometimes be seriously out of touch with the real promises and challenges of the field.Typically, this takes the form of claims that "the gene" for some complex trait—sexual orientation, for instance, or alcoholism—has been discovered. The media deserve some blame for exaggerating the significance of new research findings, but as Hewitt notes, researchers are not guilt-free: The temptation to play along with the hype in order to increase support for the field is strong.

Recent research is making such a stance increasingly untenable, however. The deeper scientists delve into the genetics of complex behaviours, the more they find that such behaviours are influenced by tens or hundreds of interacting genes, each accounting for only a small portion of the overall variance.

Other research is showing that the idea that the heritability of a given trait can be determined once and for all is mistaken. In reality, heritability for complex behavioural traits—the amount of variance in a population accounted for by genetic factors—can vary dramatically within populations.

Even those conducting animal research, which in many ways is easier to interpret than research on humans, have faced challenges. With knockout mice, for instance, developmental psychologists have been quick to point out that removing a gene from an embryonic stem cell and allowing that cell to grow into a genetically modified mouse is not the same as turning the gene off in an otherwise normal adult. The missing gene could have widespread effects on how the organism develops. Geneticists are now producing mice with conditional or inducible knockouts—genes that are inactive only during certain developmental stages, or that can be turned on or off using drugs or changes in environmental conditions. Even so, progress has been slow. Such knockouts are extremely difficult to make, she notes, and they have limitations of their own.

New techniques may help researchers overcome at least some of those challenges. One particularly promising area is the combination of behavioural genetics with visualisation tools in biology. In living animals, including humans, functional MRI and other brain-imaging techniques are providing increasingly high-resolution maps of large-scale neural activity. Meanwhile, in cells, molecular techniques such as tagging enzymes with green fluorescent protein are allowing researchers to watch changes in gene expression as they occur. Researchers are also hoping to make increasingly direct connections between animal models and clinical research. Right now, a number of interesting candidate genes have been identified in animals, but links to human behaviour are sparse.

REFERENCES

Bandura A. 1965. "Influence of models' reinforcement contingencies on the acquisition of imitative response". *J. Personal. Soc. Psychol.* 1: 589-595.

Hamilton, WD. 1964. "The genetical evolution of social behaviour I and II". *J. Theoretical Biology* 7: 1-16 and 17-52.

Robinson GE. 2002. "Genomics and integrative analyses of division of labor in honey bee colonies". *American Naturalist* 160: S160-S172.

Rothenbuhler WC. 1964. "Behaviour genetics of nest cleaning in honey bees. IV. Responses of F1 and backcross generations to disease-killed brood". *Am Zool* 4:111–123.

Scott, J.P. and Fuller, J.L. 1965. *Genetics and the Social Behaviour of the Dog*. University of Chicago Press.

4

Phylogeny and Animal Behaviour

Paleontologists are interested in understanding life through time—not just at one time in the past or present, but over long periods of past time. Before they can attempt to reconstruct the forms, functions, and lives of once-living organisms, paleontologists have to place these organisms in context. The context of evolutionary biology is phylogeny, the connections between all groups of organisms as understood by ancestor/descendant relationships. Not only is phylogeny important for understanding paleontology, but paleontology in turn contributes to phylogeny. Many groups of organisms are now extinct, and without their fossils we would not have as clear a picture of how modern life is interrelated.

WHAT IS PHYLOGENY ?

Biologists estimate that there are about 5 to 100 million species of organisms living on Earth today. Evidence from morphological, biochemical, and gene sequence data suggests that all organisms on Earth are genetically related, and the genealogical relationships of living things can be represented by a vast evolutionary tree, the Tree of Life. The Tree of Life then represents the phylogeny of organisms, i. e., the history of organismal lineages as they change through time. It implies that different species arise from previous forms via descent, and that all organisms, from the smallest microbe to the largest plants and vertebrates, are connected by the passage of genes along the branches of the phylogenetic tree that links all of Life (Figure 1).

The organisms that are alive today are but the leaves of this giant tree, and if we could trace their history back down the branches of the Tree of Life, we would encounter their ancestors, which lived thousands or millions or hundreds of millions of years ago. The notion that all of life is genetically connected via a vast phylogenetic tree is one of the most romantic notions to come out of science. How wonderful to think of the common ancestor of humans and beetles. This organism most likely was some kind of a worm. At some point this ancestral worm species divided into two separate worm species, which then divided again and again, each division (or speciation) resulting in new, independently evolving lineages. Little did these worms know, those hundreds of million years ago, that some of their number would end up evolving into beetles, while their brothers and sisters would end up as humans or giraffes.

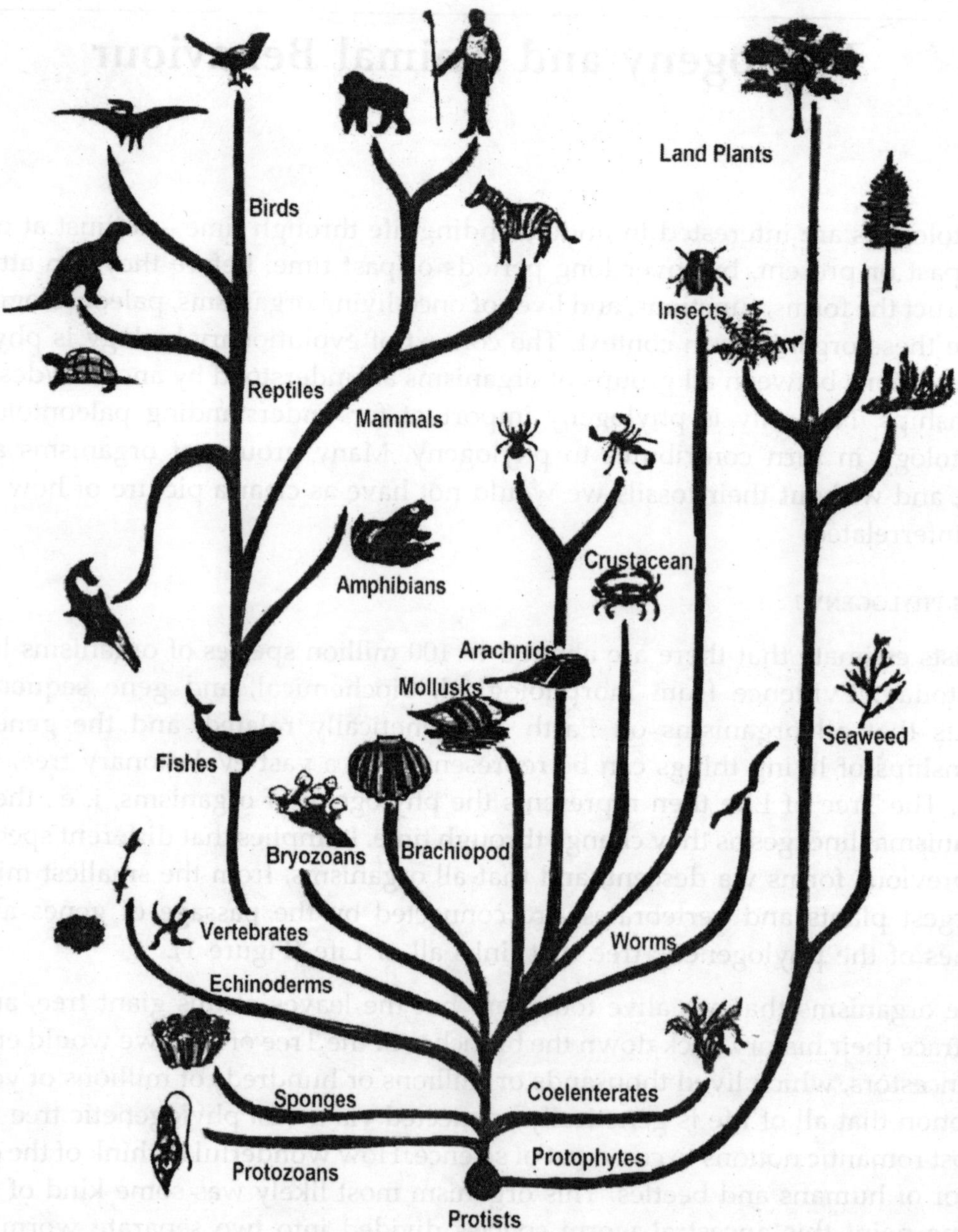

Figure 1. Phylogenetic tree of Life

Organisms have evolved through the ages from ancestral forms into more derived forms. New lineages generally retain many of their ancestral features, which are then gradually modified and supplemented with novel traits that help them to better adjust to the environment they live in. Studying the phylogeny of organisms can help us explain similarities and differences among plants, animals, and microorganisms. The Tree of Life thus provides a rigorous framework to guide research in all biological subdisciplines, and it is therefore an ideal model for the organisation of biological knowledge.

Phylogenetic Systematics

All life on Earth is united by evolutionary history; we are all evolutionary cousins — twigs on the tree of life. Phylogenetic systematics is the formal name for the field within biology that reconstructs evolutionary history and studies the patterns of relationships among organisms. Unfortunately, history is not something we can see. It has only happened once and only leaves behind clues as to what happened. Systematists use these clues to try to reconstruct evolutionary history.

A phylogeny, or evolutionary tree, represents the evolutionary relationships among a set of organisms or groups of organisms, called taxa (singular: taxon). The tips of the tree represent groups of descendent taxa (often species) and the nodes on the tree represent the common ancestors of those descendents. Two descendents that split from the same node are called sister groups. In the tree below, species A & B are sister groups — they are each other's closest relatives.

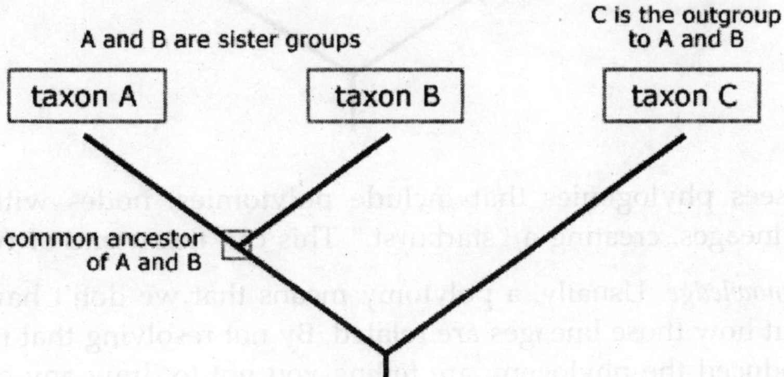

Many phylogenies also include an outgroup — a taxon outside the group of interest. All the members of the group of interest are more closely related to each other than they are to the outgroup. Hence, the outgroup stems from the base of the tree. An outgroup can give you a sense of where on the bigger tree of life the main group of organisms falls. It is also useful when constructing evolutionary trees. What's the difference between a phylogeny, an evolutionary tree, a phylogenetic tree, and a cladogram? For general purposes, not much. Many biologists, use these terms interchangeably — all of them essentially mean a tree structure that represents the evolutionary relationships within a group of organisms. The context in which the term is used will tell you more details about the representation. However, some biologists do use these words in more specific ways. To some biologists, use of the term "cladogram" emphasizes that the diagram represents a hypothesis about the actual evolutionary history of a group, while "phylogenies" represent true evolutionary history. To other biologists, "cladogram" suggests that the lengths of the branches in the diagram are arbitrary, while in a "phylogeny," the branch lengths indicate the amount of character change. The words "phylogram" and "dendrogram" are also sometimes used to mean the same sort of thing with slight variations. These vocabulary differences are subtle and are not consistently used within the biological community. The important things to remember are that organisms are related and that we can represent those relationships with tree structures.

Evolutionary trees depict clades. A clade is a group of organisms that includes an ancestor and all descendents of that ancestor. You can think of a clade as a branch on the tree of life. Some examples of clades are shown on the tree below.

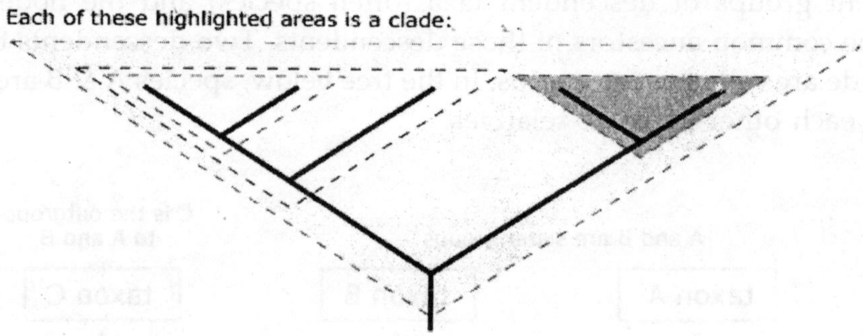

Often, one sees phylogenies that include polytomies, nodes with more than two descendent lineages, creating a "starburst." This can mean one of two things:

— *Lack of knowledge.* Usually, a polytomy means that we don't have enough data to figure out how those lineages are related. By not resolving that node, the scientists who produced the phylogeny are telling you not to draw any conclusions — and also to stay tuned: often gathering more data can resolve a polytomy.

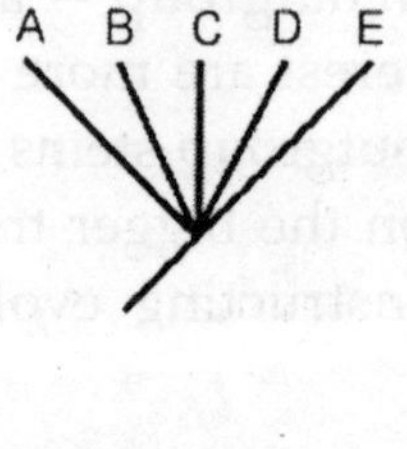

There are many ways that the polytomy above could be resolved. Six are shown below. Only more data can help us decide which is the most accurate representation of the relationships between A, B, C, D, and E.

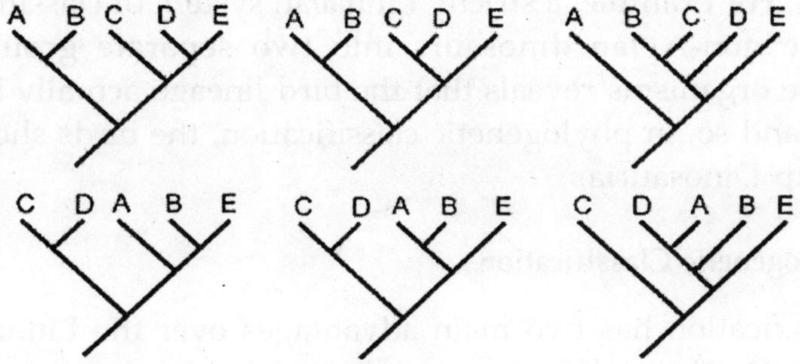

— *Rapid speciation.* Sometimes a polytomy means that multiple speciation events happened at the same time. In this case, all the daughter lineages are equally closely related to one another. The researchers who have reconstructed the tree you are examining should tell you if they feel that the evidence indicates that this is the case.

The phylogeny below shows the relationships among the members of a group of fish called cichlids. Cichlid fish speciated quickly after their home lakes formed in Africa, resulting in several phylogenetic polytomies.

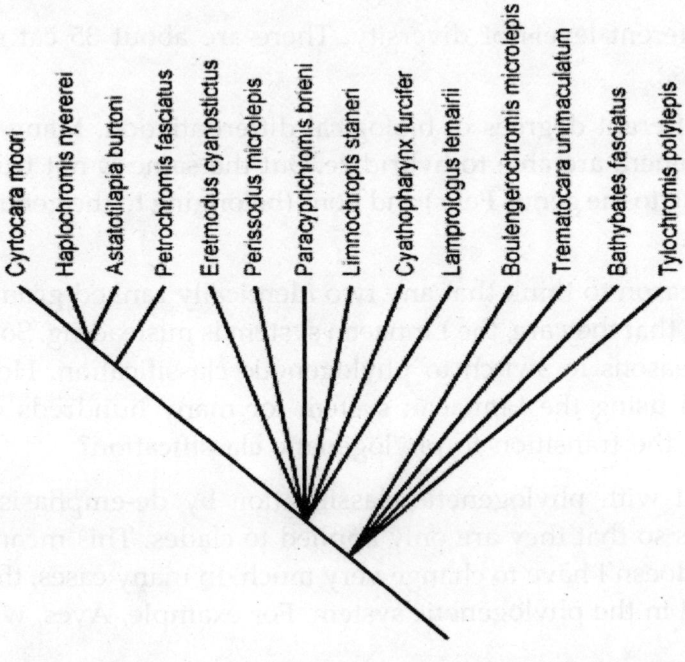

Clearly, evolutionary trees convey a lot of information about a group's evolutionary history. Biologists are taking advantage of this by using a system of phylogenetic classification, which conveys the same sort of information that is conveyed by trees. In contrast to the traditional Linnaean system of classification, phylogenetic classification names only clades. For example, a strictly Linnaean system of classification might place the birds and the non-Avian dinosaurs into two separate groups. However, the phylogeny of these organisms reveals that the bird lineage actually branches off of the dinosaur lineage, and so, in phylogenetic classification, the birds should be considered a part of the group Dinosauria.

Advantages of Phylogenetic Classification

Phylogenetic classification has two main advantages over the Linnaean system. First, phylogenetic classification tells you something important about the organism: its evolutionary history. Second, phylogenetic classification does not attempt to "rank" organisms. Linnaean classification "ranks" groups of organisms artificially into kingdoms, phyla, orders, etc. This can be misleading as it seems to suggest that different groupings with the same rank are equivalent. For example, the cats (Felidae) and the orchids (Orchidaceae) are both family level groups in Linnaean classification. However, the two groups are not comparable:

- One has a longer history than the other. The first representatives of the cat family Felidae probably lived about 30 million years ago, while the first orchids may have lived more than 100 million years ago.
- The have different levels of diversity. There are about 35 cat species and 20,000 orchid species.
- They have different degrees of biological differentiation. Many orchids belonging to different genera are able to hybridize. But the same is not true of cats — house cats (belonging to the genus Felis) and lions (belonging to the genus Panthera) cannot form hybrids.

There is just no reason to think that any two identically ranked groups are comparable and by suggesting that they are, the Linnaean system is misleading. So it seems that there are many good reasons to switch to phylogenetic classification. However, organisms have been named using the Linnaean system for many hundreds of years. How are biologists making the transition to phylogenetic classification?

Biologists deal with phylogenetic classification by de-emphasising ranks and by reassigning names so that they are only applied to clades. This means that your use of biological names doesn't have to change very much. In many cases, the Linnaean names are perfectly good in the phylogenetic system. For example, Aves, which is the class of birds in the Linnaean system, is also used as a phylogenetic name, since birds form a clade (below).

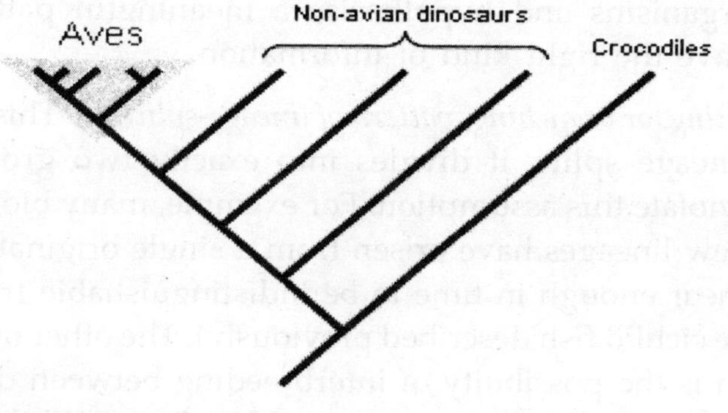

Most of the specific names that you are accustomed to using (e.g., Homo sapiens, Drosophila melanogaster) have not changed at all with the rise of phylogenetic classification. However, there are some names from Linnaean classification that do NOT work in a phylogenetic classification. For example, the reptiles do not form a clade (and cannot be a named group in the phylogenetic system) — unless you count birds as members of Reptilia too.

Cladistics

Cladistics is a method of hypothesising relationships among organisms — in other words, a method of reconstructing evolutionary trees. The basis of a cladistic analysis is data on the characters, or traits, of the organisms in which we are interested. These characters could be anatomical and physiological characteristics, behaviours, or genetic sequences.

The result of a cladistic analysis is a tree, which represents a supported hypothesis about the relationships among the organisms. However, it is important to keep in mind that the trees that come out of cladistic analyses are only as good as the data that go into them. New and better data could change the outcome of a cladistic analysis, supporting a different hypothesis about the way that the organisms are related.

There are three basic assumptions in cladistics:

1. *Change in characteristics occurs in lineages over time.* The assumption that characteristics of organisms change over time is the most important one in cladistics. It is only when characteristics change that we are able to recognise different lineages or groups. We call the "original" state of the characteristic plesiomorphic and the "changed" state apomorphic.

2. *Any group of organisms is related by descent from a common ancestor.* This assumption is supported by many lines of evidence and essentially means that all life on Earth today is related and shares a common ancestor. Because of this, we can take any collection of organisms and hypothesise a meaningful pattern of relationships, provided we have the right kind of information.

3. *There is a bifurcating, or branching, pattern of lineage-splitting.* This assumption suggests that when a lineage splits, it divides into exactly two groups. There are some situations that violate this assumption. For example, many biologists accept the idea that multiple new lineages have arisen from a single originating population at the same time, or near enough in time to be indistinguishable from such an event (as in the case of the cichlid fish described previously). The other objection raised against this assumption is the possibility of interbreeding between distinct groups, which occurs at least occasionally in some groups (like plants). While such exceptions may exist, for many groups they are relatively rare and so this assumption often holds true.

PHYLOGENETICS

Phylogenetics is the study of evolutionary relatedness among various groups of organisms (e.g., species, populations), which is discovered through molecular sequencing data and morphological data matrices. The term phylogenetics is of Greek origin from the terms phyle/phylon, meaning "tribe, race," and genetikos, meaning "relative to birth" from genesis. Taxonomy, the classification of organisms according to similarity, has been richly informed by phylogenetics but remains methodologically and logically distinct. The fields overlap however in the science of phylogenetic systematics or cladism, where only phylogenetic trees are used to delimit taxa, each representing a group of lineage-connected individuals.

Evolution is regarded as a branching process, whereby populations are altered over time and may speciate into separate branches, hybridise together, or terminate by extinction. This may be visualised as a multidimensional character-space that a population moves through over time. The problem posed by phylogenetics is that genetic data are only available for the present, and fossil records (osteometric data) are sporadic and less reliable. Our knowledge of how evolution operates is used to reconstruct the full tree.

There are some terms that describe the nature of a grouping in such trees. For instance, all birds and reptiles are believed to have descended from a single common ancestor, so this taxonomic grouping is called monophyletic. "Modern reptile" is a grouping that contains a common ancestor, but does not contain all descendents of that ancestor (birds are excluded). This is an example of a paraphyletic group. A grouping such as warm-blooded animals would include only mammals and birds and is called polyphyletic because the members of this grouping do not include the most recent common ancestor. Thus, a phylogenetic tree is based on a hypothesis of the order in which evolutionary events are assumed to have occurred.

Cladistics is the current method of choice to infer phylogenetic trees. The most commonly-used methods to infer phylogenies include parsimony, maximum likelihood, and MCMC-based Bayesian inference. Phenetics, popular in the mid-20th century but now largely obsolete, uses distance matrix-based methods to construct trees based on overall similarity, which is often assumed to approximate phylogenetic relationships. All methods depend upon an implicit or explicit mathematical model describing the evolution of characters observed in the species included, and are usually used for molecular phylogeny, wherein the characters are aligned nucleotide or amino acid sequences.

Molecular Phylogenetics

Molecular phylogenetics, also known as molecular systematics, is the use of the structure of molecules to gain information on an organism's evolutionary relationships. The result of a molecular phylogenetic analysis is expressed in a phylogenetic tree.

The theoretical frameworks for molecular systematics were laid in the 1960s in the works of Emile Zuckerkandl, Emanuel Margoliash, Linus Pauling and Walter M. Fitch.Applications of molecular systematics were pioneered by Charles G. Sibley (birds), Herbert C. Dessauer (herpetology), and Morris Goodman (primates), followed by Allan C. Wilson, Robert K. Selander, and John C. Avise (who studied various groups). Work with protein electrophoresis began around 1956. Although the results were not quantitative and did not initially improve on morphological classification, they provided tantalising hints that long-held notions of the classifications of birds, for example, needed substantial revision. In the period of 1974–1986, DNA-DNA hybridisation was the dominant technique.

Every living organism contains DNA, RNA, and proteins. Closely related organisms generally have a high degree of agreement in the molecular structure of these substances, while the molecules of organisms distantly related usually show a pattern of dissimilarity. Conserved sequences such mitochondrial DNA are expected to accumulate mutations over time, and assuming a constant rate of mutation provide a molecular clock for dating divergence. Molecular phylogeny uses such data to build a "relationship tree" that shows the probable evolution of various organisms. Not until recent decades, however, has it been possible to isolate and identify these molecular structures.

The most common approach is the comparison of sequences for genes using sequence alignment techniques to identify similarity. Another application of molecular phylogeny is in DNA barcoding, where the species of an individual organism is identified using small sections of mitochondrial DNA. Another application of the techniques that make this possible can be seen in the very limited field of human genetics, such as the ever more popular use of genetic testing to determine a child's paternity, as well as the emergence of a new branch of criminal forensics focused on evidence known as genetic fingerprinting.

The effect on traditional biological classification schemes in the biological sciences has been dramatic as well. Work that was once immensely labor- and materials-intensive can now be done quickly and easily, leading to yet another source of information becoming available for systematic and taxonomic appraisal. This particular kind of data has become so popular that taxonomical schemes based solely on molecular data may be encountered.

Early attempts at molecular systematics were also termed as chemotaxonomy and made use of proteins, enzymes, carbohydrates and other molecules which were separated and characterised using techniques such as chromatography. These have been largely replaced in recent times by DNA sequencing which produces the exact sequences of nucleotides or bases in either DNA or RNA segments extracted using different techniques. These are generally considered superior for evolutionary studies since the actions of evolution are ultimately reflected in the genetic sequences. At present it is still a long and expensive process to sequence the entire DNA of an organism (its genome), and this has been done for only a few species. However it is quite feasible to determine the sequence of a defined area of a particular chromosome. Typical molecular systematic analyses require the sequencing of around 1000 base pairs. At any location within such a sequence, the bases found in a given position may vary between organisms. The particular sequence found in a given organism is referred to as its haplotype. In principle, since there are four base types, with 1000 base pairs, we could have 41000 distinct haplotypes. However, for organisms within a particular species or in a group of related species, it has been found empirically that only a minority of sites show any variation at all and most of the variations that are found are correlated, so that the number of distinct haplotypes that are found is relatively small.

In a molecular systematic analysis, the haplotypes are determined for a defined area of genetic material; ideally a substantial sample of individuals of the target species or other taxon are used however many current studies are based on single individuals. Haplotypes of individuals of closely related, but supposedly different, taxa are also determined. Finally, haplotypes from a smaller number of individuals from a definitely different taxon are determined: these are referred to as an out group. The base sequences for the haplotypes are then compared. In the simplest case, the difference between two haplotypes is assessed by counting the number of locations where they have different bases: this is referred to as the number of substitutions (other kinds of differences between haplotypes can also occur, for example the insertion of a section of nucleic acid in one haplotype that is not present in another). Usually the difference between organisms is re-expressed as a percentage divergence, by dividing the number of substitutions by the number of base pairs analysed: the hope is that this measure will be independent of the location and length of the section of DNA that is sequenced.

An older and superseded approach was to determine the divergences between the genotypes of individuals by DNA-DNA hybridisation. The advantage claimed for using hybridisation rather than gene sequencing was that it was based on the entire genotype, rather than on particular sections of DNA. Modern sequence comparison techniques overcome this objection by the use of multiple sequences.

Once the divergences between all pairs of samples have been determined, the resulting triangular matrix of differences is submitted to some form of statistical cluster analysis, and the resulting dendrogram is examined in order to see whether the samples cluster in the way that would be expected from current ideas about the taxonomy of the group, or not. Any group of haplotypes that are all more similar to one another than any of them is to any other haplotype may be said to constitute a clade. Statistical techniques such as bootstrapping and jackknifing help in providing reliability estimates for the positions of haplotypes within the evolutionary trees.

This example illustrates several characteristics of molecular systematics and its underlying assumptions.

— Molecular systematics is an essentially cladistic approach: it assumes that classification must correspond to phylogenetic descent, and that all valid taxa must be monophyletic.

— Molecular systematics often uses the molecular clock assumption that quantitative similarity of genotype is a sufficient measure of the recency of genetic divergence. Particularly in relation to speciation, this assumption could be wrong if either some relatively small genotypic modification acted to prevent interbreeding between two groups of organisms, or in different subgroups of the organisms being considered, genetic modification proceeded at different rates.

In animals, it is often convenient to use mitochondrial DNA for molecular systematic analysis. However, because in mammals mitochondria are inherited only from the mother, this is not fully satisfactory, because inheritance in the paternal line might not be detected: in the example above, Vilà et al. cite more limited studies with chromosomal DNA that support their conclusions.

These characteristics and assumptions are not wholly uncontroversial among biological systematists. As a cladistic method, molecular systematics is open to the same criticisms as cladistics in general. It can also be argued that it is a mistake to replace a classification based on visible and ecologically relevant characteristics by one based on genetic details that may not even be expressed in the phenotype. However the molecular approach to systematics, and its underlying assumptions, are gaining increasing acceptance. As gene sequencing becomes easier and cheaper, molecular systematics is being applied to more and more groups, and in some cases is leading to radical revisions of accepted taxonomies.

Ernst Haeckel's Recapitulation Theory

During the late 19th century, Ernst Haeckel's recapitulation theory, or biogenetic law, was widely accepted. This theory was often expressed as "ontogeny recapitulates phylogeny", i.e. the development of an organism exactly mirrors the evolutionary development of the species. Haeckel's early version of this hypothesis [that the embryo mirrors adult evolutionary ancestors] has since been rejected, and the hypothesis amended as the embryo's development mirroring embryos of its evolutionary ancestors. Most modern biologists recognise numerous connections between ontogeny and phylogeny, explain them using evolutionary theory, or view them as supporting evidence for that theory. Donald Williamson suggested that larvae and embryos represented adults in other taxa that have been transferred by hybridisation (the larval transfer theory).

Gene transfer

In general, organisms can inherit genes in two ways: vertical gene transfer and horizontal gene transfer. Vertical gene transfer is the passage of genes from parent to offspring, and horizontal gene transfer or lateral gene transfer occurs when genes jump between unrelated organisms, a common phenomenon in prokaryotes. Lateral gene transfer has complicated the determination of phylogenies of organisms, since inconsistencies have been reported depending on the gene chosen. Carl Woese came up with the three-domain theory of life (eubacteria, archaea and eukaryotes) based on his discovery that the genes encoding ribosomal RNA are ancient and distributed over all lineages of life with little or no lateral gene transfer. Therefore rRNA are commonly recommended as molecular clocks for reconstructing phylogenies. This has been particularly useful for the phylogeny of microorganisms, to which the species concept does not apply and which are too morphologically simple to be classified based on phenotypic traits.

Taxon Sampling and Phylogenetic Signal

Owing to the development of advanced sequencing techniques in molecular biology, it has become feasible to gather large amounts of data (DNA or amino acid sequences) to estimate phylogenies. For example, it is not rare to find studies with character matrices based on whole mitochondrial genomes. However, it has been proposed that it is more important to increase the number of taxa in the matrix than to increase the number of characters, because the more taxa the more robust is the resulting phylogeny. This is partly due to the breaking up of long branches. It has been argued that this is an important reason to incorporate data from fossils into phylogenies where possible. Using simulations, Derrick Zwickl and Hillis found that increasing taxon sampling in phylogenetic inference has a positive effect on the accuracy of phylogenetic analyses. Another important factor that affects the accuracy of tree reconstruction is whether the data analysed actually contain a useful phylogenetic signal, a term that is used generally to denote whether related organisms tend to resemble each other with respect to their genetic material or phenotypic traits.

PHYLOGENY AND ONTOGENY OF AFFECTIVE SOCIAL BEHAVIOUR

The Galapagos Islands finches have been used by Charles Darwin and many other scientist to study how " random variation and natural selection can drive the production of organisms with novel features, adapted to new ways of life. The finches studied in the Galapagos Island are usually characterised by the great variance in beaks and the function of beaks. The finches beaks were first thought to change over many years because of local ecological condition, but many scientist devised a new hypothesis which stated, "As a consequence of beak evolution, there have been changes in the structure if finch vocal signals." This hypothesis was based the assumption that beaks have a direct correlation in effecting song production.

The Galapagos Island finches displayed a wide variety phenotypes of beaks with different forms and functions caused by natural selection over many years. This phenotype variation was hypothesised by the Grants to be a result of natural selection because of ecological conditions. The new studies, however, tested to see if finch beaks have an affect on the acoustic structure of the songbird songs. The first experiment done to test this hypothesis involved first checking to see if beak movement indeed caused sound production, which in fact turned out to be true. Thus, one could imply that beak variations could cause a different song production.

The goal of these researches is " to show how research on the relationship between beaks and song is providing novel insights into the interplay of morphological adaptation and the evolution communication signals." The bird song is actually made in the birds syrinx, which is similar to the human larynx during speech production. The scientist then addressed the question of how the birds are able to send pure tonal signals across such a long range and distance. The answer is that the birds are able to change their vocal tracts configurations, which in turn directly changes the frequency of their songs.

The three generalisations, which these finches fall into describing their songs are, simple, variability, and cross-species overlap. If the beak is opened more wide then it creates a higher pitched sound. So it could be hypothesised that a large beaked bird will have a large vocal tract and will in turn create a lower song frequency. This hypothesis was backed up by the studies of Maria Palacious and Pablo Tubáro who tested the Neotropical Woodcreepers Pendrocolaptinae, which produced lower frequency songs with long beaks. These scientists tested, these highly protected finches, by video taping them and then determining the beak gapes for each song for the different finches.

The two main reason male finches use their singing, like many other singing birds, is for territory defense and mate attraction, but the difference in beak size could affect both these characteristics, because some males with either small or large beaks could not produce the right song in other to perform these tasks. This lead Nowick and colleagues to test to see if a beak changed its size for food collection, then would this have a direct affect on how fast or slow a finch could move its beak to produce its specific song. Nowick then concluded that there was a positive correlation between "divergences of beaks and its influence not only the feeding but also their singing behaviour." The scientist, however, plan to broaden their investigation by studying how wider beaks may influence songs and also study other birds with beak morphology such as the crossbills or the Hawaiian honeycreepers.

Human consciousness and uniquely human behavioural distinctives are the end products of a phylogenetic history shared with other hominids, plus a lengthy and uniquely human ontogenetic development. Human behavioural distinctions include the ability to infer abstract causality, abstract theory construction, language embodied in culture, morality, artistic endeavour and spirituality.

Human distinctives arise as a result of emotion-based relationships with other persons. A vast literature exists on the necessity of a social setting for the development of language, but other cognitive abilities are also clearly dependent upon early learning which in a human infant necessitates relationship with a caregiver. The development of an embodied spirituality is another human distinctive that can only emerge through nurturant interaction with other persons, both human and the divine.

Emotion plays a major role in the emergence of these human attributes by providing the valenced states that motivate development of neural circuitry, and this role is especially prominent during early development when the brain is most plastic. The cognitive evolution of the human infant occurred in parallel with the evolution of a uniquely human brain, characterised by an expanded neocortex, extensive lateralisation and cytologically unique neurons such as the von Economo or spindle cells, nevertheless, even these specific developments were built on neural adaptations that humans share with other mammals.

Primary Emotions

Long term genetic selection has resulted in the existence in vertebrates of basic neural programmes providing valenced affective states which prepare and guide responses to appropriate stimuli. Primary emotions provide the core nature of the value system guiding the refinement of synaptic connections in interaction with the physical and social environment, hence primary emotions provide the emotional palette that guides brain development. This idea is an elaboration of Panksepp's formulation of affective neuroscience, which describes how neurobiological systems mediate the basic emotions.

Affective neuroscience could be considered a compliment to Edelman's theory of neural Darwinism, dealing with how brain development and function can be well understood in terms of a process of natural selection acting on neural connections. In the case of "neural Darwinism" the time-course of the selection process referred to is not that of the millennia of phylogeny but rather the developmental course of an individual brain.

The term "neural Darwinism" is used in this context to stress the idea of survival during the development of synaptic connections of the strongest, most effective and repeatedly used circuits. Extensive synaptic pruning and apoptosis occur during the early development of each individual brain as excess neural circuits are initially formed and then as individual behaviours are tuned by environmental influence the neurons and synapses which are not of utility do not survive. This means that the basic neural patterns are somewhat "hard-wired" into the developing early brain with the caveat that extensive fine-tuning of neural circuitry takes place predominantly during early stages, but continuing at a declining rate throughout life.

Primary emotional pathways are laid down early and at subcortical limbic and brainstem levels and will tend to be more "hard-wired" than later-forming neocortical pathways and incidentally will be less accessible to conscious awareness. This means that primary emotions, although subject to some fine-tuning by environmental and social stimulation will tend to be more influenced by genetic control and less plastic than cognitive abilities. The primary emotional systems thus come to underlie the later development in the individual of intellectual capacities. Primary emotions also underlie the more behaviourally flexible secondary emotions which arise from blending of primary emotions or interaction with learning.

Basic sensory-motor modules involved in pattern-recognition, motor output and problem-solving have access to both the emotional and cognitive systems and are also necessary to enable development of neural circuitry. Through the actions of neurotransmitter, neuromodulator and neurotrophin levels, the wiring of brain circuitry and thereby intellect and self-awareness is widely influenced by the nature, timing and amount of activation in the emotional circuits during early life. Dopamine release in the ventral tegmental pathway is one specific example of how this occurs.

The release of brain derived neurotrophic factor (BDNF) as pathways are exercised is another example. Higher brain functions are sculpted on the basis of these primary emotions providing the valenced states that select which neurons, synapses and pathways will survive during subsequent maturation of the brain. It follows that elucidation of the specific nature of these systems is crucial to understanding the way the brain functions and structures itself.

In a work which virtually inaugurated the discipline of affective neuroscience, Jaak Panksepp described seven "primitive emotional operating systems that exist in limbic and reptilian areas of the brain". He designated these as the SEEKING, RAGE, FEAR, LUST, PANIC (separation distress), CARE, and PLAY systems. These are hierarchically organised executive operating systems which give rise to valenced affective states during interaction with several layers of non-specific perceptual, attentional, and cognitive processes. In this view primary emotions are action promoting valenced states with distinct neural circuitry and neurochemistry the consequences of which outlast the precipitating conditions.

In contrast reflexive affects such as hunger, pain, and sensory affects such as taste or smell are closely time-locked to their triggering stimuli and are not primary emotions. Panksepp stresses that the differences between primary or prototypical emotions and secondary emotions, such as shame and guilt, include instantiation in more ancient medial and ventral brainstem pathways which are richer in innervation from the viscera and utilise a variety of visceral neuropeptides.

Primary emotional systems are those with a long phylogenetic history and they organise complex but flexible reactions by activating or inhibiting autonomic, hormonal and/or somatic changes that were adaptive during evolutionary history. Cosmides and Tooby also point out that an emotion is a superordinate programme which orchestrates and integrates the activities of various functional subprogrammes, which include reflexive affects, and also subprogrammes governing perception, cognitive appraisals and feeling states. The specific combination of behavioural components will depend on context and eliciting stimuli.

Basic Functioning and Survival of Individual

Panksepp's SEEKING System is the primary task-oriented pathway by which affective goals are met. It is activated on the one hand by primary biological needs characterised by homeostatic signals, but also by signals from other primary and secondary emotional systems and conscious volitional desires. The SEEKING system energises activity on the basis of a perception of need or dissatisfaction. It is based in the mesolimbic dopamine pathway, but extends to orbitofrontal cortex in humans. It is generalised in its goals able to be activated by any specific need but can also function in a non-specific manner.

One is a motivational, appetitive system and the other an hedonic appraisal system linked to consummatory activities and satisfaction. The distinction is illustrated by the fact that addiction involves craving of substances or experiences but not necessarily experiencing satisfaction by them. The wanting and liking components seem to be separate entities. These two systems can be behaviourally dissociated and function independently in rats and in humans. Therefore the hedonic appraisal component should be considered separately from the SEEKING system and proposed the existence of a separate PLEASURE system in addition to Panksepp's SEEKING system.

Primary emotions of social bonding

The emotional systems promoting social bonding are LUST, PLAY, CARE, and PANIC. Sexual reproduction, essential for long-term survival of the species is the outcome of the sexual LUST system, clearly an ancient adaptation found in all vertebrates. With the development of mammals there were significant additions to the basic vertebrate social repertoire. Social bonding is however supported by the sexual attraction system and vasopressin and/or oxytocin, hormones involved in sexual behaviour also facilitate pair bonding and are involved in other social behaviours.

According to MacLean the basic differences between reptiles and the later evolving mammals are,

1) lactation and maternal care,
2) vocal communication to maintain mother-infant contact, and
3) playful behaviour facilitating social learning.

By definition lactation and maternal care are essential for the survival of mammals and these would have been subject to significant selection pressure. Compared with other mammals, hominid infants have a relatively long period of helplessness combined with subsequent need for training in foraging techniques and social behaviours. This is true of all hominids but two additional factors discussed by Falk contributed to the extended helpless period of human infants. The first was the trend toward a narrow pelvis associated with bipedalism, and the second was the expansion of the human brain. Together these factors would have selected for human infants that were delivered at an increasingly immature stage of development. Because nurture is essential for survival of immature infants, and the infancy of hominids became increasingly extended there must have been selection pressure for establishment of neural circuitry for emotional attachment between mother and infant. This circuitry would be expected to be most highly developed in humans.

In higher animals, social bonding and group cohesion is initially effected primarily by the NEED/ATTACHMENT or Separation Distress System in the young, which triggers panic if there is separation distress, but which also provides satisfaction during closeness. It was necessary for the complementary CARE system , through which parents respond to the young, to evolve in tandem.

The biological origins of human sadness are rooted in the extended brain system involving the cingulate gyrus that mediates this separation distress in infant animals although this neurological substrate has an even longer evolutionary history in that it was used earlier in phylogeny for perception of physical pain. This is perhaps the reason why humans perceive separation from loved ones as so very similar to pain.

Infant chimps lack the ability to cling to their mothers' fur in the first two months and so must be supported by the mother on her ventral surface. Mothers of older infants use body language and gestural signals to encourage climbing on the mother's back when it is time to move on. The mother's use of gestures and facial expressions play a key role in communication with infants and infants develop an intense interest in their mother's face. Chimp mothers also teach which foods can be eaten and perhaps even tool use. Chimp infants signal distress to the mother through whisper, hoo and scream vocalisations. Hominid evolution therefore involved tandem evolution of emotional circuitry in adults to provide not merely food, also but emotional nurturance and instruction, and the parallel circuitry in the young to seek and respond to caring adults. This entailed the increasing use of gesture, facial expression, tactile and vocal communication.

The increasingly extended period during which hominin young were dependent on caregivers necessitated the development of even more skilled caretaking and the ability of adults to provide instruction. Although the role of these emotions in producing adult human emotional and social behaviour has been well studied, the role of emotion as a selection factor in cognitive development has been less well researched. Mother-infant communication likely provided the emotional motivation for the initial development of language, and its use in adult coalitions was likely a more secondary development.

Learning in human infants has been shown in numerous studies to be critically enabled by reciprocal interaction with the primary care-giver in the early stages of life. The ability for shared attention between infant and mother is critical for the development of a theory of other minds and influences the development of language. Conversely childhood neglect or separation from the caretaker leads to developmental stunting, as for example in hospitalisation syndrome, and appears to be correlated with intellectual and/or social impairment in later life.This phenomenon has been observed in other mammals, but is particularly striking in primates. Thus social emotions provide the valenced state necessary for infant learning, initially taking place in relation to predicting and responding to the actions and emotions of the primary carer.

These systems give rise to the adult need to be part of a social group and tendencies to respond to others in a caring manner. Studies in voles have identified oxytocin as being implicated both with maternal and adult pair-bonding. That social bonding is supported in part by the more ancient circuitry of the sexual attraction system is suggested in that both vasopressin and/or oxytocin, hormones involved in sexual behaviour, facilitate other forms of social bonding. These systems also facilitate altruistic behaviour in animals, especially social animals such as cetaceans, canids and primates which have relatively well developed memories for social interactions.

Learning, the crucial basis of all higher development, is enabled by the SEEKING system but in young organisms strongly facilitated by the CARE and NEED systems in tandem with the PLAY System. The tendency for young mammals to be involved in play as part of their preparation for both food procurement and adult social roles, suggests that play should also be considered a basic emotional programme in the human ancestral lineage and necessary for the normal cognitive development of children. The PLAY system develops significant cortical components in higher mammals parallel with its importance in learning. Chimp and bonobo mothers in particular engage in extensive nuzzling, tickling, play-biting, chasing and other forms of play with their infants and play periods are accompanied by facial gestures and vocalisations often including laughter.

In mammals, play involves learning social roles and social behaviours. It is facilitated evolutionarily by the enlargement of the cerebral cortex and the prolonged infant/ maternal interaction necessitated by lactation. During play social commitment may be revealed involuntarily and assessed by others. Playful teasing allows the exploring of boundaries. These social aspects may provide one basis for the evolution of altruistic behaviour, and these social aspects also enable enhanced learning. There is a real possibility that optimal cortical plasticity may depend on the activation of play, affection, and other rewards of close attachment. Allowing juvenile rats 30 minutes of rough and tumble play results in increased BDNF transcription in the amygdala and dorsolateral frontal cortex.

Although much anecdotal information suggests this may also be true in humans the underlying neurophysiological mechanisms are still poorly defined. Play is particularly important in language development. It is also an essential component of performing arts, ceremonial and celebratory behaviour and an important source of creativity. The phylogenetic transition from rough and tumble play alone to the capacity for imaginative play facilitated the development of language through the necessity of understanding and empathising with others that is a critical part of imaginative play.

For many species, group living is crucial for survival advantage, both in terms of finding food and protection against predators as well as in enabling learning. However this inevitably entails a competition for resources that needs regulation to minimise damage to individuals as well as group cohesion. Allocation of rank occurs in animals and humans alike by various competitive processes leading to agonic behaviour that regulates competition in a socially non-destructive way.

Competition takes place in relation to the control of "territory" regarded in the widest sense, relating not merely to material resources (such as food, material possessions, and geographical areas) but also to social control, sexual mates, status symbols, and intellectual turf. Identity is closely related to territory regarded in this sense. In humans this becomes embodied in social roles and associated status, and underlies many social activities such as competitive sport, and is fundamental in many social arrangements. Thus it becomes a central part of cultural systems. Sewards and Sewards describe a "power dominance" neural system in mammals which gives rise in humans to subjective feelings of self-esteem, the need to excel, to succeed and to overcome obstacles.

It is a very old vertebrate system which functioned to allow competition for mates, territories and other material resources and later gave rise to competition for status and social approval. The striatal complex, which includes the basal ganglia, constitutes a major portion of the "reptilian brain" which is deemed by MacLean to control instinctive, phylogenetically older behavioural patterns. The striatal system referred to by MacLean as a "reptilian" component has continued to evolve in mammals playing a major role in motor behaviours. Additional limbic structures including the anterior cingulate gyrus, which plays a role in so many social behaviours, were recruited in mammals for this system. Dominance thus came to involve well developed strategies in hominids who had the intellectual ability for planning and the memory to keep track of subtle interactions and alliances.

The POWER/dominance System is the affective means employed by the evolutionary process to regulate potentially destructive competition for resources by enabling social dominance of some individuals while allowing others to survive and to wait for better opportunities. It involves a desire in humans for higher rank but also an acceptance under normal conditions of assigned status. The associated feelings are pride/high self-esteem in satisfactory circumstances, and shame/low self-esteem when they are unsatisfactory.

The POWER system involves anterior cingulate cortex (ACC) plus non cortical areas such as peri-aqueductal gray, and hypothalamus. Association with the LUST system and competition for mates is indicated by the fact that vasopressin and testosterone are likely involved in its regulation. Furthermore alterations in both vasopressin levels and ACC activity are involved in depression, a disorder that evolutionary psychiatrists see as the human counterpart of social defeat. The development of neocortex in humans allowed for complex integration with cognition resulting in emergence of secondary emotions such as guilt, shame and jealousy. Its malfunctioning can lead to psychiatric disorders the existence of which shed light both on the normal functioning of the system and its evolutionary origins . There is much evidence that low status is a major risk factor for depression in mammals, as this probably indexes low probability that a host of needs will be met.

Human Distinctives

The human distinctives of theory of mind, language, extensive and creative culture, morality and spirituality all developed in parallel, but language evolution likely played a central role in enabling the emergence of the other attributes. These complex developments in humans depend on the existence of several primary emotional systems. Beings endowed with a single emotional system—merely a feeling of well-being or unhappiness—would not be capable of developing the same range of higher-level repertoires, including moral behaviour. One result of this is that specific psychological or psychiatric problems are likely to arise if specific primary emotional systems are dysfunctional, for whatever reason. Recognising this feature may be useful in terms of social and medical care. Another result is that the emergence of the full range of human potential depended on the phylogenetic development of emotion-based values.

THEORY-OF-MIND

We believe that the evolutionary development of a theory of other minds was linked specifically to the evolution of the ability to recognise emotions in the self and others. Chimps and bonobos and perhaps some other apes may possess the rudiments of a theory-of-mind as they are able to anticipate the needs and intentions of others and modify their own behaviours accordingly. This has been well documented in the literature for dominance interactions among adults and it is also clearly a part of maternal infant interaction.

More significantly, enculturated apes such as the bonobo Kanzi are able to pass linguistically mediated theory-of-mind tests, showing that the neurological prerequisites for theory of mind are available to bonobos. Perhaps ontological development of theory-of-mind requires involvement with caring others who also possess this ability, for example through joint attention which forms during mother infant interactions, and it then becomes socially mediated through linguistic interaction. At any rate ability to be aware of self and others paralleled the development, not of sensory cortex, but of the emotionally salient frontal and limbic cortex.

Interoceptive functions, as well as the experience of disgust, are based on the anterior insula (AI) which in primates possesses an input pathway allowing it (rather than lower structures such as the amygdala and striatopallidum) to become the major focus for valuation of stimuli. The AI and ACC are on one hand architecturally and phylogenetically more primitive than sensory neocortex, but on the other hand , in humans, chimps, bonobos and dolphins they contain aunique recently evolved cellular type (the von Economo (VEN) or spindle neuron. VENs are hypothesised to be involved in expectancy of behavioural outcomes, human social intuition and formation of a theory of mind. Interestingly self-awareness only arises in humans, great apes, and perhaps some cetaceans i.e. those animals with von Economo neurons. The number of these cells in each species is correlated across both species and age of individuals with degree of self-awareness. Consistent with a proposed role for VENs in self-awareness, humans attain their full compliment of VENs only at about 4 yrs of age.

Core representations of the body arising from structures such as the insula may play a role not only in self-awareness, but also in an "as-if" loop system that allows evaluation and anticipation of events without their actual occurrence. The somatic marker hypothesis proposes that the insula provides sensory representation of the state of the body as part of this "as-if" loop mechanism.

In support of this, Goldman and Sripada have proposed that the reading of basic emotions in the faces of others might utilise specialised neural programmes, which evaluate emotions of others based on simulation of that state in oneself. During memory this may be accomplished by a reverse simulation which would require the type of "as-if"loop proposed by Damasio. The same "as if" circuitry might function in mirroring the bodily and emotional states of others in the manner in which actions are mirrored by motor neurons in primate brains. As noted above the experience of disgust and its recognition in others activates the AI and ACC.

The experience of pain in self and others also activates these areas and response to pain in others is even modulated by the perceived fairness of that person suggesting a neural substrate for empathy. Activation of this region in response to facial expressions of others has been shown recently to correlate with self-reported the ability to empathise with others. The functional role of VENs is as yet unknown, but it would be consistent if they were involved in the neural simulations, which make empathy and theory of mind possible.

Damage to the insula and adjacent frontal area produces deficits in moral behaviour, which exist in spite of normal intellectual processing VENs in the frontoinsular and ACC areas are specifically damaged in frontotemporal dementia, a disorder in which emotional and social awareness, empathy and theory of mind are disrupted. Additionally some autistic individuals who have deficits in the theory-of-mind in others have been shown to have dysfunction in these same cells.

Thus the phylogenetic appearance of VENs is associated with the development of empathy, theory of mind and the ability for moral reasoning. This development might be considered to be a preadaptation influenced by the role of the insula in awareness of bodily state in self and others—a role originally arising from the need to evaluate and avoid contact with and contamination by noxious substances. The brain area originally involved in visceral sensations and evaluating "good taste" and "bad taste" became a necessary component of the neural circuitry for empathy, judging fairness, and making moral decisions.

LANGUAGE

The evolution of language ability required not only the evolution of complex conceptual structures, and systems of rules to encode them, but as many have noted some motivation to represent and communicate concepts. Why did language develop only in one species? Recent suggestions for the initial selection pressure have stressed adult social interactions such as a Machiavellian motive to outwit group members, gossip as a substitute for grooming, or the need for males to provide social displays to attract females or the perhaps the need to regroup after scavenging. Bickerton argues against these social theories on the basis of the fact that other primate troops (e.g. baboons) can be very large and complex, but there seemed to have been no selection pressure for other primates to develop anything approaching protolanguage.

The relative ease with which children develop language suggests a relation between the structure of language and its acquisition. This led to Chomsky's proposal of the existence of a Universal Grammar. One possible explanation for the appearance in all human lineages of a universal grammar is that special brain mechanisms slowly evolved to support language.

A second theory is that brain mechanisms supporting language appeared suddenly perhaps due to a small number of mutations. The discovery of a role of the FOXP2 gene in language spurred a great deal of initial excitation for those in the latter camp. A variation of this theme that language is the result of a faculty unique to humans is held by Fitch, Hauser and Chomsky who see the essential features of language as exaptations from other abilities.

Another possibility suggested by Christiansen and Chater is that rather than the brain adapting to support language providing a neural basis for universal grammar, language itself was the subject of selection pressure. Language was selected for ease of learning by hominins who already had particular neural substrates. Because language adaptation is culturally based, changes in language due to cultural evolution occur much faster than biological changes in the brain. One might expect that any so-called "language genes" that appeared in a population would diverge especially rapidly, as early humans spread out geographically, leaving the question of how an original universal grammar remained universal. Therefore, it is more likely as Christiansen and Chater suggest that language evolved to fit pre-existing structures of the brain rather than the other way around, and language was adapted to use a general purpose rather than a specialised modular strategy.

The structures that were available for the use of language were the limbic structures used by primates in vocalisation, in particular the ACC, as well as the motor structures used in gesturing. It has been suggested that because the analogue of Broca's area in the monkey contains mirror neurons responsive to gestures in other individuals, language arose through gradual adaptation of the gestural system. Bonobos use gestures in a flexible manner in communication with conspecifics adjusting for the attentional state of the recipient. Subadults are the most common users of gestures and social learning appears to be responsible for acquisition of at least some specific gestures.

Non-human primate vocalisations are involuntary because they arise from limbic rather than neocortical areas whereas human speech is under the voluntary control of a motor area, namely Broca's area.Captive common chimpansees, however, have been shown to use vocalisation in a voluntary manner, not associated with emotional outburst, to direct attention of onlookers. Gaining attention from others is then in these chimps both voluntary and a socially motivated behaviour. The enculturated bonobo Kanzi's use of spoken English indicates that it is neurologically possible for bonobos to vocalise voluntarily. Perhaps the cingulate gyrus which initiates vocalisation in monkeys provides the emotional flavour in bonobos for voluntary vocalisation through its known connections with motor areas.

Even chimp mothers and infants take turns during play perhaps providing a basis for the development in protomotherese of a more voluntary action than that assumed for typical primate vocalisations. Kanzi has been shown to engage in the type of communicative turn taking that suggests a pattern of human discourse even though the individual segments of discourse are very short and devoid of complex syntax. Analysis of his discourse with trusted caretakers suggests that he is indeed well aware of the communicative uses of discourse with vocal and gestural components. The way in which Kanzi and his kin learned language, however, was quite different from previous attempts at teaching language to apes. These bonobos have been exposed to language early in life in a manner approximating that of human children within a family setting. They have clear emotional ties to their human caregivers and exist as part of a unique Pan/Homo culture. The relationship has not been merely social, it is social with strong emotional ties, attachment and nurture. Some of the videos of caregivers interacting with Kanzi suggest that they use language with inflections and vocal stress similar to motherese. The nurturing atmosphere and "family" like upbringing may in fact be the key to ape language learning.

Modern human motherese uses exaggerated vowels, hyperarticulation, heightened prosody, and exaggerated facial expressions all of which contribute to emotion regulation in infants. Human infants in turn are predisposed to respond to motherese which then prepares the infant to be attuned to the phonetic and semantic aspects of their native language. Motherese also maintains joint attention and encourages infants to learn turn taking aspects of dialogue.

The emotional aspects of motherese cause it to be so attractive to infants, but the end result is that the emotional motivation leads to the ontological process of learning phonological and semantic aspects of language. Other modern hominids do not vocalise nearly as much with infants, but the evolutionary trend toward lengthened childhood, immature infants and increasing brain size were likely coupled with an increased selection pressure for more complex nurturing of infants which in turn led to increasing joint attention, understanding of other minds, and use of communicative gestures and sounds.

The voluntary use of nonsense sounds in mother-infant play would be motivated by the social emotion of CARE and PLAY, but freed from the more immediate and reflexive nature of e.g. alarm calls. Representational play and chasing may have been extended at that point to include actions feigning intentions such as "I'm going to get you", but these were freed from aggressive intent. As sounds also began to be used in play, siblings might begin to use in common some of the same sounds used by their mother, and also use these sounds in play with other unrelated juveniles. In this manner local cultures of sound use might develop. Rather than the development of a universal protolanguage, it would be enough for further evolution of language that some individuals learned to use sounds or gestures generally to represent concepts. Over long periods of time, local cultures of juveniles may then have developed their own proto-dialects though cultural learning but motivated by social play. The concept that one thing can stand for another would have been the necessary development which emerged in part from representational and feigning play.

Proto-motherese and proto-language likely had multimodal origins in a structureless combination of gesture, facial expression, gaze, vocal and non-vocal sounds. In addition other emotional needs such as those involved in mating and social grooming probably worked concurrently to provide multiple selection pressures on the use of protolanguage. It is probably significant that Broca's area and its primate analogue contain mirror neurons for both hand and facial movements and this probably lead to the use of both types of signals concurrently. Over time, protolanguage would come to rely more heavily on the vocal mode although modern speech is still strongly connected to gesture.

Admittedly there remain some problems with any continualist theory, including ours. If protolanguage was holistic with strings of utterance rather than employing short discrete units, a mechanism is needed to segment longer utterances into meaningful individual units (words). The fractionation of longer holistic strings into short segments might be an intellectually more difficult task than learning individual words as symbols. This presents a problem for the theory that language evolved from holistic protolanguage. However even the use of a very few short sounds may have led to the realisation that sounds may be representational. The evidence for representational play in chimps, and pretense in captive bonobos, indicates that this ability was available to early hominins as well.

Another problem that might be raised is how a fully developed syntax could emerge from protolanguage consisting of something like the telegraphic speech used by modern two year olds. The evolution of the ability to represent recursion is viewed by most as a necessary component of the evolution of language. Recent experiments in which starlings were able to discriminate syntactic patterns, however, suggest that perhaps rudimentary recursion is not an exclusively human ability. In addition there is perhaps one modern language, Piraha, which lacks the use of recursion, as well as colour names, numeracy, and past and future tenses and this case has been used to challenge Chomsky's model of universal grammar. Of course, the case of Piraha does not tell us whether recursion is necessary for language to evolve, as its loss in this group may be a secondary development. A recent hypothesis put forward by Okanoya is that the unique components necessary for language, in particular recursion, emerged from interactions among more general preadaptations for emotion, sensory motor-integration, and a mirror system of imitation.

MORALITY

At present, an extensive debate exists around the extent to which innate mental modules influences human moral behaviour. Some biologists such as Marc Hauser propose the existence of dedicated moral modules similar to Chomsky's language modules. However, in light of the fact that it is now agreed that several emotions seem to contribute to moral behaviour, the idea of a dedicated moral module is questionable. As with the evolution of language, it is difficult to see how all the various preadaptations would come together into one genetically controlled module. Jonathan Haidt has in contrast proposed the existence of several evolutionarily prepared moral domains each of which offers survival benefits in a social setting and which act somewhat independently. These include 1) altruism and kinship ties 2) reciprocity, and concern for fairness 3) rank and authority 4) in-group loyalty, and 5) moral purity and avoidance of contamination.

Each of the emotional operating systems has the potential to either overtly or subliminally influence behaviour thereby providing motivational force for Haidt's suggested moral domains. These emotional operating systems are part of the human biological heritage and influence the content of our moral behaviours. Because each emotional system has been moulded by unique selection pressures and uses distinctive neural circuitry the influence each has on moral intuition remains separate and independent of the influence of other emotions.

The emotional systems that have been adapted to function during social encounters play obvious roles. The primary emotional POWER system influences both the authority/subordination domain described by Haidt and the domain giving rise to reciprocity/fairness/concern for justice. In addition, the CARE nurturance system of adults plays an obvious role in moral intuitions and together with the separation distress (PANIC) system has been adapted to support various forms of altruistic behaviour. The reciprocity and fairness domain of Haidt may derive in part from the PLAY system, which promotes learning of appropriate social behaviours in juveniles.

Schaller has proposed what he calls an "emotional immune system" based on an evolutionary bias to avoid others who might be diseased, but which may extend to other areas of social avoidance, cultural attitudes and religious rituals. Humans avoid others with obvious disease but also any difference from the norm—including asymmetries, disabilities and obesity—because difference from the norm may signal underlying ill health or poor genetic potential. Foreigners are also more likely to carry pathogens for which we have no immunity. Schaller's theory thus bridges the disease- avoidance function of disgust with emotional prohibitions concerning touch and sexual contact with others. Morals are biologically biased to consist of rules for sexual contact and social proximity to avoid biological contamination.

HUMAN CULTURE

Although culture has been shown to exist in animals the explosion of human creativity that seemed to occur only within the last 100,000 years was likely linked to language evolution. Human imagination artistic endeavours, and various aspects of culture have roots in the emotional programming of young mammals, and primates in particular, to engage in play behaviours with conspecifics, a behaviour which functions to underpin learning and preparation for adult life.

The expanded human brain, extended childhood, and cultural scaffolding provided during childhood by older individuals are all necessary for the development of human nature. Kanzi and other enculturated apes are high achievers precisely because they have been provided with the cognitive scaffolding of their human caretakers. Their unique language, imitation of intentional actions and toolmaking abilities would not have emerged without emotional involvement and shared attention with humans, which enabled learning of behaviours that they did not invent themselves.

Furthermore culture is a progressive process as those who have lived and worked with these apes note that the young born to already enculturated bonobos "gaze, think, gesture and behave in nonverbal ways that increasingly diverge" from non-enculturated bonobos. Although even wild apes engage in representational play, only those bonobos with acquired linguistic ability engage in play which involves true pretense showing that not only does cultural scaffolding further language use, but language also furthers cultural development.

REFERENCES

Bateman AJ. 1948. *Intra-sexual selection in Drosophila.* Heredity 2: 349-368.

Carew TJ. *Behavioural Neurobiology: The cellular organisation of natural behaviour,* Sinauer Assoc. Inc. Sunderland, Massachusetts.

Darwin CR. 1871. *The Descent of Man and Selection in Relation to Sex,* Hurst and Co.

Hamilton, WD. 1964. "The genetical evolution of social behaviour I and II". *J. Theoretical Biology* 7: 1-16 and 17-52.

Scott, J.P. and Fuller, J.L. 1965. *Genetics and the Social Behaviour of the Dog.* University of Chicago Press.

Wilson EO. 2000, *Sociobiology: The New Synthesis,* Twenty-fifth Anniversary Edition, Belknap Press.

5

Animal Behaviour and Cognition

Animals behave, and they perform sophisticated information processing. The cognitive capabilities such as visual scene analysis, motor behaviour, and problem solving, etc., are found in non-humans as well as humans. Research on chimpanzees-who share over 95% of our DNA sequences-shows that they engage in many complex mental and emotional processes resembling those of humans. Animals are also part of the history of American psychology. Psychologists have been studying and testing them for well over 100 years, and such famous psychologists as Watson, Thorndike, and Skinner worked primarily with animals.

Historically, the study of animal behaviour is interesting because there are two distinct threads-the American approach called comparative psychology and the European approach called ethology-that came together in the middle of the 20th Century. First they were antagonistic, then they formed a fruitful synthesis, to mutual benefit.

Meanwhile, the technological advances of recent decades (such as brain scanning and microminiaturisation of electronics) have greatly increased the sophistication of research on animal cognition. Those advances, along with progress in understanding genetics, has allowed modern psychologists to place human intelligence in a more accurate context, highlighting our relationship to what came before as well as pinpointing exceptional and unique human talents.

Animal cognition is the title given to a modern approach to the mental capacities of non-human animals. It has developed out of comparative psychology, but has also been strongly influenced by the approach of ethology, behavioural ecology, and evolutionary psychology. The alternative name cognitive ethology is therefore sometimes used; and much of what used to be considered under the title of animal intelligence is now thought of under this heading. In practice, animal cognition mostly concerns mammals, especially primates, cetaceans and elephants, besides canidae, felidae and rodents, but research also extends to non-mammalian vertebrates such as birds such as pigeons, lizards or fish, and even to non-vertebrates (cephalopods).

In 1907, Clever Hans, a horse that was claimed to have been able to perform arithmetic and other intellectual tasks made the New York Times. After formal investigation in 1907, psychologist Oskar Pfungst demonstrated that the horse was not actually performing these mental tasks, but was watching the reaction of his human observers. The horse was responding directly to involuntary cues in the body language of the human trainer, who had the faculties to solve each problem. The trainer was entirely unaware that he was providing such cues.

For most of the twentieth century, the dominant approach to animal psychology was to use experiments on intelligence in animals to uncover simple processes (such as classical conditioning and operant conditioning) that might then account for the apparently more complex intellectual abilities of humans. This approach is well summarised in the mid-century book by Hilgard, but its reductionist philosophy was combined with a strongly behaviourist methodology, in which overt behaviour was taken as the only valid data for the study of psychology, and in its more extreme forms (the radical behaviourism of B. F. Skinner and his experimental analysis of behaviour) behaviour was taken as the only topic of interest. In effect, the mental processes that humans experience in themselves were viewed as epiphenomena .

The success of cognitive psychology in addressing human mental processes, which began in the late 1950s and was proclaimed by Neisser, led to a re-evaluation of the research paradigm, and researchers began to address animal mental processes from the opposite direction, by taking what is known about human mental processes and looking for evidence of comparable processes in other species. In a sense this was a return to the approach of Darwin's protegé George Romanes, arguably the first comparative psychologist of the modern era. However, whereas Romanes relied heavily on anecdote and an anthropomorphic projection of human capacities onto other species, modern researchers in animal cognition are in most cases firmly behaviourist in methodology, even though they differ sharply from the behaviourist philosophy. There are some exceptions to the rule of behaviourist methodology, such as John Lilly and, some would argue, Donald Griffin, who have been prepared to take a strong position that other animals do have minds and that humans should approach the study of their cognition accordingly. However, their claims have not found wide acceptance in the scientific community, though they have attracted an enthusiastic following among lay people.

The development of animal cognition was also strongly influenced by:

- increased use of and interest in primates (and also cetaceans) rather than the rats and pigeons that had become the classic species of the comparative psychology laboratory, and by developments within primatology;
- advancing knowledge of animals' behaviour in their natural environments through studies in ethology, sociobiology and behavioural ecology; such studies often

showed that animals needed certain cognitive abilities in order to adapt to their ecological niche;

— one or two high profile projects, in particular Allen and Beatrice Gardner's Washoe project in which a chimpanzee learned at least some elements of American Sign Language.

— advancing understanding of brain function through work in physiological psychology and cognitive neuropsychology

This account of the history of the study of animal cognition is inevitably oversimplified. From Romanes on, there have always been comparative psychologists who have been more or less cognitively inclined: obvious examples are Wolfgang Köhler, famous for his studies of insight in chimpanzees, and Edward C. Tolman, who introduced into psychology, as an explanation of the behaviour of rats in mazes, two ideas that have been immensely influential in human cognitive psychology - the cognitive map and the idea of decision-making in risky choice according to expected value.

ANIMAL COGNITION : THEORETICAL ISSUES

Animal cognition is the study of the processes used to generate adaptive or flexible behaviour in different animal species. As a part of cognitive science, research in animal cognition aims to uncover the different cognitive mechanisms at play across species in order to understand the varieties of cognition, the similarities between humans and other species, and the evolution of cognitive processes.

As with human cognition, there are competing theories about the structure of animal cognition at play. After the cognitive revolution many animal cognition researchers believed that cognition is a general purpose computational system. This was especially true of those who advocated the Social Intelligence Hypothesis. According to the hypothesis, the relatively sophisticated general problem-solving capacities of social animals are due to challenges arising from social living. This view has been challenged by those who claim that the research on other species suggests that cognition is more modular than the traditional view allows. Some argue that individuals cannot use one mechanism to solve all cognitive tasks, since there are different rates of learning and different computational processes implicated in different domains. Because of these differences, cognition must be thought of as a number of special-purpose computational modules rather than one general purpose processor. Other researchers examine animal cognition from within the framework of embodied cognition, or dynamical systems theory.

Animal cognition research has historically accepted a close relationship between affect and cognition. Early experimental psychologists manipulated subjects' motivation (e.g. by withholding food) and the pain associated with shocks were assumed to cause an unpleasant affective response. Today there is growing interest in the emotional responses of animals. In this sense, animal cognition research has anticipated the recent interest in emotion in cognitive science.

Philosophical discussion about animal cognition has traditionally focused on the metaphysics and epistemology of mind in creatures who do not have language. While the traditional debate about the existence of animal minds is problematic given the lack of clarity about the nature of mind, recent discussions of animal beliefs and rationality help to make the discussion less muddy.

The early history of western philosophy reflects a tendency to see animals as lacking rationality. Aristotle defined "human" as "the rational animal", thus rejecting the possibility that any other species is rational (Aristotle Metaphysics). Aquinas believed that animals are irrational because they are not free (Aquinas Summa Theologica). Centuries later Descartes defended a distinction between humans and animals based on the belief that language is a necessary condition for mind; on his view animals are soulless machines (Descartes Discourse on the Method). Locke agreed that animals cannot think, because words are necessary for comprehending universals (Locke Essay Concerning Human Understanding). Following in this tradition, Kant concludes that since they cannot think about themselves, animals are not rational agents and hence have only instrumental value.

However, there were dissenters. Voltaire criticised Descartes' view that humans but not animals have souls and hence minds, by suggesting there is no evidence for the claim (Voltaire Philosophical Dictionary). Hume was downright dismissive of the animal mind skeptics when he wrote "Next to the ridicule of denying an evident truth, is that of taking much pains to defend it; and no truth appears to me more evident than that beasts are endowed with thought and reason as well as man. The arguments are in this case so obvious, that they never escape the most stupid and ignorant"

Animal Minds

Despite Hume's judgment about their worth, much ink has been spilled developing arguments for the existence of animal minds. The two standard arguments are extensions of the two arguments for other minds: the argument from analogy and the inference to the best explanation argument. The argument from analogy for animal minds can be formulated as:

— All animals I already know to have a mind (i.e. humans) have property x.

— Individuals of species y have property x.

— Therefore, individual of species y probably have a mind.

Versions of the argument differ as to what they choose as the reference property x, and how they defend the choice of reference property. The reference property might refer to some general capacity (e.g. problem solving), some specific ability (e.g. language, theory of mind, tool-use), or it might consist of some set of properties.

The argument from analogy for animal minds is in one sense stronger than the argument for other minds insofar as the reference class is larger; rather than starting with the introspection of one's own mind and then generalising to all other humans, the argument for animal minds takes as given that all humans have minds and generalizes from the human species to other species. In another sense, the analogical argument for animal minds is weaker, since the strength of the argument is a function of the degree of similarity between the reference class and the target class. Humans might be more similar to one another than they are to members of another species.

The second standard argument for animal minds rests on the claim that the existence of animal minds is a better explanation of animal behaviour and physiology than those offered by other hypotheses. This argument can be formulated as:

— Individuals of species x engage in behaviours y.

— The best scientific explanation for an individual engaging in behaviours y is that it has a mind.

— Therefore, it is likely that individuals of species x have minds.

The inference to the best explanation argument justifies the attribution of mental states to animals based on the robust predictive and explanatory power that is gained from such attributions. As the argument goes, without such attributions we would be unable to make sense of animal behaviour. This argument makes use of standard methods of scientific reasoning; of two hypotheses, the one that better accounts for the phenomenon is the one to be preferred. Those who offer this sort of argument for animal minds are claiming that behaviourist or other mechanistic explanations for animal behaviour fail to account for the diversity and flexibility of behaviour in at least some species of animal. Critics of this argument offer alternative explanations for the relevant behaviours.

Rationality, Beliefs, and Concepts

Though philosophers and psychologists seem to generally accept that other species have minds, there is widespread disagreement about what exactly that means. For some, it merely means that some species can feel pain (and hence there are codes of conduct for working with animals). For others, it means that some species are rational agents who have reasons for action. The theoretical investigation into animal cognition examines whether any other species are rational, have beliefs, or have concepts. There are corresponding epistemological questions about how we might ever know the content of an animal's belief or reason for action, given the difficulty of attributing content to a creature who doesn't use language. Whether animals are rational is related to philosophical questions having to do with the moral status of animals. For example, arguments in favor of personhood for great apes are made on the basis of the rational capacities of these species, as are arguments for duties toward animals, and arguments given in favor of moral agency (or proto-moral agency).

Discussions of animal rationality are confounded by the lack of consensus on what is required for rationality. Because there are many different kinds of rationality (e.g. practical vs. theoretical, process vs. product), disagreements about what sorts of cognitive mechanisms are implicated in rationality (e.g. linguistic processing, logical reasoning, causal reasoning, simulation, biases and heuristics), and disagreements about the extent to which different kinds of normativity are implicated in rationality (e.g. biological fitness or reason-respecting propositional attitudes), there is no straightforward way to answer the question about whether other species are rational agents. Some philosophical theories of rationality are based on an initial acceptance of rationality across species, given evolutionary considerations. For example, on Fred Dretske's view, even some simple learned behaviours, such as a bird's avoiding eating a monarch butterfly, can be construed as minimally rational. Because monarchs who eat toxic milkweed become toxic to birds and other predators, when a bird learns not to eat monarch butterflies after having formed an association between eating monarchs and vomiting, it has a reason for its avoidance behaviour. The birds also have a reason to avoid eating a viceroy, given that it is visually almost indistinguishable from a monarch, though not poisonous. The behaviour in both cases is explained by the content of the bird's thought (or "thought"), and for Dretske this is sufficient for the bird to count as a minimally rational agent.

Given the lack of theoretical consensus on the nature of rationality, empirical research projects are not designed to examine rationality directly. Instead, researchers investigate various capabilities that may be associated with rationality. For example, tool-use has long been considered to be an indicator of rational thought. Because tool use involves finding or constructing an object that is utilised as an extension of the body to achieve a goal, it is thought that tool use requires identifying a problem, considering ways of solving the problem, and realising that other objects can be used in the manipulation of the situation. Early experimental research on chimpanzee problem solving by the German psychologist Wolfgang Köhler had chimps constructing tools to acquire out of reach objects; it was reported that chimpanzees would stack boxes or put together tubes to form a long rod in order to reach bananas hung overhead. Given this behaviour, Köhler suggested that chimpanzees solve some problems not by trial-and-error or stimulus response association, but through a flash of insight. Since the days of Köhler, tool use in the wild has been discovered in a number of different taxa, including all great apes, some monkeys, some birds, sea otters, and cetaceans.

Some theoretical arguments about animal rationality identify rationality with other properties, such as having beliefs or concepts. Donald Davidson has offered an argument against animal rationality based on an association between concepts, beliefs, and language. On Davidson's view, believers must have the concept of belief, because to have a belief they must recognise that beliefs can be true or false, and one cannot understand objective truth without understanding the nature of beliefs. In order to develop an understanding of objective truth, one must be able to triangulate with others, to talk to others about the world, and hence all believers must be language users. Since other species lack language, they do not have beliefs. Davidson also argues against animal beliefs based on the claim that having a notion of error is necessary for being a believer.

A different argument against animal belief has been presented by Stephen Stich, who argues that we cannot attribute propositional attitudes to animals in any metaphysically robust sense, given our inability to attribute content to an animal's purported belief. On Stich's view, if attribution of belief to animals is understood purely instrumentally, then animals have beliefs. However, if attribution of beliefs to animals requires that we can accurately describe the content of those beliefs, then animals don't have beliefs. Given the second sense of having belief, Stich argues that because "nothing we could discover would enable us to attribute content to an animal's belief", we are unable to make de dicto attributions to other species, and we cannot make de re attributions because this would violate the truth-preserving role of attribution. Hence we can make no attribution, and if we can't say what an animal's belief is about, it makes no sense to say that an animal has a belief. The worry here is similar to the worry about anthropomorphism; when we use our language to ascribe content to other species, we may be attributing to them more than is appropriate. Stich is concerned that when we say "Fido believes there is a meaty bone buried in the backyard" we are attributing to Fido concepts he cannot possibility have, concepts like "backyard" which are only comprehensible if one has corresponding concepts such as "property line", "house", "fence", and so on. Stich's argument can be formulated as:

— In order for something to have a belief, it must have a concept.

— In order to have a concept, one must have particular kinds of knowledge, including knowledge of how the concept relates to other concepts.

— Non-human animals don't have this knowledge.

— Therefore, non-human animals don't have beliefs.

While animal cognition researchers agree that we ought to be careful about the concepts that we attribute to other species, many deny Stich's claim that empirical research cannot help us learn anything about the conceptual organisation of other species. One of the earliest attempts to examine animal concepts came out of a series of experiments with pigeons. The subjects were shown photographs, and were trained to peck at the pictures that contained a target object, such as a tree, and not respond to the pictures that didn't contain the target object. After the training period, the pigeons were able to generalise to new photographs, pecking only at those that contained trees just as in the training set. It was suggested that this sorting ability demonstrates that the pigeon has a concept of the target object. Many reject the idea that being able to sort objects is sufficient for having a concept corresponding to the sortals. For one, some think language is necessary for concept acquisition. Others think that while concept acquisition is independent of language, sorting behaviour alone doesn't demonstrate having a concept, because humans can be trained to sort objects while lacking the corresponding concept. As Colin Allen and Marc Hauser put it, "It is possible to teach a human being to sort distributors from other parts of car engines based on a family resemblance between shapes of distributors. But this ability would not be enough for us to want to say that the person has the concept of a distributor".

Rather than identifying concept acquisition with sorting behaviour, Allen and Hauser suggest alternative methodologies for identifying concepts in other species. For example, they offer a possible (though, they admit, ethically untenable) test for a death concept among vervet monkeys. Vervet mothers are capable of recognising the alarm cries of their infants, and when they hear such a cry the mother will look towards her infant, and the other females will look towards the mother. Allen and Hauser suggest that playing a recording of a recently deceased infant's alarm cry would help to determine whether vervets have a concept of death. If the mother responds to these recordings in an atypical fashion, unlike the usual response made to a living infant, that response provides evidence of a death concept. According to Allen and Hauser, having a concept permits different responses to identical stimuli. The actual sound of the infant's alarm cry during life is identical to the sound played back after death. If the response to this stimulus is different, this is evidence that there has been a conceptual change associated with the stimulus. Allen presents the general strategy for attributing concepts to animals as follows: "An organism O may reasonably be attributed a concept of X (e.g. TREE) whenever:

- O systematically discriminates some Xs from some non-Xs; and
- O is capable of detecting some of its own discrimination errors between Xs and non-Xs; and
- O is capable of learning to better discriminate Xs from non-Xs as a consequence of its capacity".

One way to study the conceptual structure of other species is to use the same methods that are used to study concepts in another group that lacks language, namely human infants. The preferential looking time paradigm, also known as the dishabituation paradigm, is used to study human infants' understanding of the physical and social world. Dishabituation experiments are thought to help us understand what kinds of predictions infants make about their word, and this information can help us determine how they see the world. The methodology is simple; an infant is repeatedly shown a stimulus, and after becoming habituated to the stimulus the infant becomes disinterested. At this point, a new stimulus is shown. If the infant sees the new stimulus as different from the target stimulus, or impossible given the target stimulus, the infant will look longer at the new stimulus. If the infant takes the new stimulus to be similar to the target stimulus, then she will not show any additional interest. The idea is that by comparing responses among groups of individuals, a researcher can learn something about how that group conceptualizes the world.

In one study using this method, Marc Hauser and colleagues investigated numerical concepts in different primate species, including rhesus monkeys (Hauser et al. 1996) and cotton-top tamarins. The researchers tested the monkeys' ability to keep track of individual objects placed behind a barrier. They found that like human infants, the monkeys look longer at impossible outcomes. For example, in one test condition the rhesus monkeys were shown two eggplants serially placed behind a screen, and then the screen was lifted showing only one eggplant. The monkeys looked longer at the one eggplant than they did when shown the expected two eggplants, suggesting that they represent the eggplants as distinct sortals.

Another way we might learn how different species organise the world is to teach individuals a symbolic communication system. For example, the biologist Irene Pepperberg trained an African Grey parrot named Alex to sort objects using meta-level concepts that categorise other concepts. Alex was able to sort objects according to color, shape, and matter, and he was able to sort sets of objects according to number. In addition to sorting, Alex could report which feature makes two objects similar or different. For example, when presented with a red block and a red key, Alex responded to the question "What's same?" by uttering "color." He could also report similarities and differences in shape and matter. Pepperberg claims that her studies demonstrate Alex's understanding of categorical concepts, and reveal the classifications that Alex devised. However, one might be worried that the concepts exhibited by symbol-trained animals are an artifact of the communication system, and not typical of the species.

Anthropomorphism

When researchers attribute mental states to other species, they open themselves to the charge of anthropomorphism. The term "anthropomorphism" refers to the act of attributing uniquely human traits to other animals; the traits in question are usually psychological states. In recent years, there have been a number of theoretical discussions about the charge of anthropomorphism itself.

In response to charges that psychological and agential attributions are examples of anthropomorphic attribution, some have argued that the charge of anthropomorphism is a charge that the attributor is making a category mistake, rather than merely a false attribution. It is a claim that the attribution must be logically false, because members of the target species are not the sorts of things to which the term can apply. However, if the charge of anthropomorphism is the charge that the attributer is making a category mistake, then the charge is being made on conceptual, rather than empirical grounds; hence the worry that refusing to attribute so-called anthropomorphic properties without first examining whether they might be held by members of any other species is unscientific. Thus, one response to the charge of anthropomorphism is continued research, for one wouldn't know whether a property is anthropomorphic until after the relevant research has been completed. As Sober puts it, "The only prophylactic we need is empiricism".

However, Sober also argues that the empirical methodology of psychology places a different burden of proof on animal cognition and human cognition research. This is because the null hypothesis in the animal cognition research is that there are different cognitive mechanisms at work in humans and animals. Given that type 1 errors (reporting a false positive and rejecting a (possibly true) null hypothesis) are taken to be more serious errors than are type 2 errors (reporting a false negative and not rejecting a null hypothesis when it should be rejected), the practice of science results in a bias against attributing psychological traits to animals. The debate about how to interpret the results of animal studies as compared to human studies may be seen as a debate about an inconsistent application of Morgan's Canon. Morgan's Canon states: "In no case may we interpret an action as the outcome of the exercise of a higher psychical faculty, if it can be interpreted as the outcome of the exercise of one which stands lower in the psychological scale. Though this is a longstanding rule of thumb in animal cognition research, sometimes referred to as the "principle of conservatism," it is not a principle commonly used in human cognition research. To complicate matters, attempts to determine what exactly Morgan's Canon instructs a researcher to do have raised worries about its meaningfulness.

Despite the defenses given for attributing mental states to animals, worries about anthropomorphism remain. Kennedy claims that the arguments for attributing mental properties to animals often rest on a false dichotomy: either animals are stimulus-response machines, or they are agents with beliefs and desires. Since animals are not stimulus-response machines, they must be psychological agents. The problem with this argument is that not all machines implement stimulus-response functions; some machines are complex and indeterministic, and if animals were machines, they would be machines of that sort.

Other critics rely on arguments much like those of Stich and Davidson discussed above. If we have no good scientific methods for attributing mental states to creatures without language, then we should not make such attributions. Since we are barred from making the attributions, scientific psychology ought not engage in analysing animal mentality. Anthropomorphism is seen as a human tendency that must be overcome in order to do good science.

It has been noted that such arguments are about the proper methods of science, the scope of science, and how to interpret data. As such, the argument is not an empirical one, but a theoretical one. This can be seen in the way the debates sometimes result in an impasse. Those opposed to attributing mental properties to animals are accused of begging the question, by committing "reverse anthropocentrism" or "anthropodenial" . The charge of begging the question goes both ways. Kennedy argues that arguments for animal mentality are grounded in human intuition or introspection and that introspection is itself anthropomorphic and ought not be taken as objective evidence. As both sides accuse the other of begging the question, some conclude that the debate is not fecund, and ought to be replaced with empirical work in comparative biology and psychology.

The concerns about anthropomorphism appear to be largely limited to western scientists. It has been argued that researchers from countries with a Buddhist rather than Christian orientation are not culturally encouraged to see a categorical distinction between humans and nonhuman animals. Unlike Christianity, Buddhist doctrine does not claim that humans but not animals have immortal souls, and it does not allow humans to use animals for their own purposes in the ways Christianity does. The Buddhist tradition sees a connection between humans and other animals, and allows that humans can be reborn as animals. De Waal argues that the difference in cultural attitudes toward animals led to an early rejection of Japanese methods and findings in primatology, and that it is only recently that some of those ideas, such as Kinji Imanishi's claim that primates display cultural differences within species, have made their way into western scientific discourse.

THE SCIENCE OF ANIMAL COGNITION

Scientific interest in animal minds and cognitive capacities grew as a result of Charles Darwin's theory of evolution by natural selection. In The Descent of Man, Darwin introduced many of the issues that motivate the research programmes in animal cognition today, including tool use, reasoning, learning, concepts, consciousness, the social sense, and the moral sense. He was also interested in animals' aesthetic judgments and whether they believe in the supernatural, issues that haven't been taken up by contemporary researchers. In addition, Darwin anticipated current interest in implicit reasoning with his comment "The savage would certainly neither know nor care by what law the desired movements were effected; yet his act would be guided by a rude process of reasoning, as surely as would a philosopher in his longest chain of deductions"

Darwin advocated the continuity of the mental across species; just as some morphological characteristics are homologous across species living in similar environments, we should expect psychological and behavioural similarities as well: "the difference in mind between man and the higher animals, great as it is, certainly is one of degree and not of kind". This view was also advocated by Darwin's contemporary, the naturalist George Romanes, who in his book Animal Intelligence writes "there must be a psychological, no less than a physiological, continuity extending throughout the length and breadth of the animal kingdom".

Anecdotal Method

The method that Darwin, Romanes, and other contemporaries used to investigate these questions could be described as the anecdotal method. Stories about animal behaviour were collected from a variety of people, including military officers, amateur naturalists, and layfolk, and were compiled and used as evidence for a particular cognitive capacity in that species.

The anecdotal method as practiced by Darwin and Romanes has been criticised for a number of reasons. The "evidence" gathered was often a story told about an event witnessed by a single person, usually not a trained scientific observer. In addition, these stories were often acquired second- or third- hand, so there were worries that the reports had been embellished or otherwise altered along the way. These problems were recognised early on, and in response Romanes developed three principles for accepting anecdotes in order to avoid some of these problems:

- Never accept an incident report as fact without considering the authority or respectability of the observer.
- If the observer isn't known, and the incident report is sufficiently important, consider whether the observer may have reason or cause to make an inaccurate report.
- Look for corroborations of the observation by examining similar or analogous observations made by other independent observers.

The third principle was the one he most relied on, writing "This principle I have found to be a great use in guiding my selection of instances, for where statements of fact which present nothing intrinsically improbable are found to be unconsciously confirmed by different observers, they have as good a right to be deemed trustworthy as statements which stand on the single authority of a known observer, and I have found the former to be at least as abundant as the latter".

Despite Romanes' attempts, the method remained problematic insofar as it didn't provide any statistical information about the frequency of such behaviours; selection bias would lead people to report only the interesting intelligent behaviours and ignore the frequency of behaviours that might serve as counterevidence. Thus, the anecdotal method as practiced by Darwin and Romanes fails many of the virtues associated with good scientific method.

The legacy of Darwin and Romanes for animal cognition can be summed up in the oft-quoted phrase "The plural of 'anecdote' is not 'data'." Though 19th century anecdotalism has been rejected, defenses of the use of observations (often now referred to as "incidents" or "qualitative reports" rather than "anecdotes") come from ethology. Among ethologists, collecting behavioural incidents is part of the standard methodology, and fieldwork by trained scientists has resulted in greater knowledge of animal behaviour. For example, when Jane Goodall reported seeing chimpanzee hunting and lethal intergroup aggression, the scientific image of chimpanzees had to be significantly revised.

A middle ground in the use of qualitative reports takes the collection of incidents as a research tool. To be useful as a research tool, the concerns about anecdotalism must be addressed. One major concern has to do with the interpretation and description of the behaviour as it is reported. Like concerns about attributing content to animal minds, the worry is that there will be an over-attribution due to the bias of the scientist. In addition, if anecdotes are presented as stories, and the story structure is used to determine truth, there is a worry that such stories will be elaborated so as to remove inconsistencies or fill gaps. The psychologist Richard Byrne defends the scientific use of rare events as a useful tool, writing, "careful and unbiased recording of unanticipated or rare events, followed by collation and an attempt at systematic analysis, cannot be harmful. At worst, the exercise will be superseded and made redundant by methods that give greater control; at best, the collated data may become important to theory". It is important to note that Byrne is not talking just about data acquired by the ethological method, but also incidents observed by scientist in the field or lab who are well-versed in the baseline behaviour of the species.

Experimental Method

Before Watson and Skinner promoted behaviourism in human psychology, similar ideas were being espoused among animal cognition researchers who were disappointed in the lack of rigorous methods for studying animal minds. The psychologist C. Lloyd Morgan's early text on comparative cognition was critical of the anthropomorphism in the anecdotes reported by Darwin, Romanes, and others. In An Introduction to Comparative Psychology, Morgan famously proposed what is now called Morgan's Canon; recall that Morgan's Canon is a principle of conservatism that instructs researchers to interpret behaviour as caused by the lowest possible "psychical faculty". In his text Morgan suggested that many of the seemingly cognitively sophisticated behaviours of animals could be explained by associative learning.

The Clever Hans scandal of 1904 demonstrated Morgan's Canon in use; Hans was a famous Russian trotting horse who charmed crowds by appearing to calculate mathematical problems, as well as to read German and musical notation, simply by tapping his hoof. After much investigation, the experimental psychologist Otto Pfungst found that Hans wasn't counting or reading language; rather he was reading his owner von Osten's bodily motions. Von Osten was unconsciously cuing Hans to start and stop tapping his foot at the correct time, and Hans had merely leaned to associate von Osten's movements with the correct behaviour. Today, the legacy of Clever Hans can be seen in the controls used during experimental tests of an animal's ability. For example, researchers who know the correct response will wear a welder's mask, blackened goggles, or some other device to keep the subject from being cued by eye gaze or facial expressions, or naive trainers are used during testing.

One of the leading animal cognition researchers during this time was Edward L. Thorndike (1874-1949). Thorndike argued for the necessity of experimental study of animal intelligence, writing, "most of the books do not give us a psychology, but rather a eulogy of animals. They have all been about animal intelligence, never about animal stupidity". Experiments will help us learn both what animals can do and what they fail to do, thus giving us a better overall understanding of animal cognition. One of Thorndike's projects focused on the problem solving abilities of housecats. He placed cats in a variety of puzzle boxes, and observed the strategies the cats used to escape. When first put in a new box, the cats took a long time to find the solution, but after experience with the box they were able to escape much more quickly. Thorndike found that the cats improved by ignoring the ineffective actions and performing the useful ones. This suggested to Thorndike that cats learn through trial and error, and his conclusion helped to reinforce the belief that animal behaviour can be fully explained in associative terms.

While the behaviourists succeeded in introducing much-needed rigor into the study of animal minds, there was some concern that they had gone too far, that the methods were too stringent, and that the drive for repeatable and controlled experiments could not be used to uncover all there is to know about the function of animal minds. For example, the ethologists thought that to understand animal behaviour, animals must be observed in their natural environment. As sterile laboratory experiments stripped of social and environmental context, some considered the behaviourists' studies ecologically invalid. While today some experimentalists defend these methods, other experimentalists agree with the criticisms, and in response have developed methods that are sensitive not only to environmental concerns, but also to development and social relations. For example, research coming out of Kyoto University's Primate Research Institute (PRI) is based on a three-part research programme. First, chimpanzee physical, cognitive, and social development is taken into account in the design of experiments, and subjects are raised by their mothers rather than by human caregivers or unrelated animals. In addition, lab work and fieldwork is synthesised; field observations are used to develop experiments, and experiments are conducted both in the field and in the laboratory. Finally, the method includes analysis of the physiological and biological features of the species that could be related to cognitive abilities.

The experimental research at PRI uses what they call the "participant observation" method, which is based on the triadic social relationship between mother, infant, and experimenter. When testing chimpanzees in the lab, they are never taken from their natural social environment; rather, the experiments are brought into the social environment. As a researcher becomes a member of that social environment, she can run experiments that are woven into normal daily activities. At PRI, a different researcher is bonded with each mother-infant dyad, and the relationship is expected to last a lifetime. This close relationship between human and chimpanzee is thought to offer many benefits. It makes the chimpanzees more willing to engage in the research activities, so researcher can better understand of what chimpanzees can and cannot do (rather than what they will and won't do). In addition, Matsuzawa claims that the participant observation method is better at investigating species typical social cognition than are isolated experiments on single subjects, because the PRI chimpanzees are not integrated into a human social environment, but the researchers adapt to the chimpanzee social environment. Finally, the bond between researcher and subject allows the human to interact with his chimpanzee "research partners" at a younger age, given the trust between researcher and mother. Mother and infant can be taught a task together, which can help to illuminate developmental differences in particular abilities. For example, Inoue and Matsuzawa have recently reported that infant chimpanzees are better able to recall strings of numerals in order than are adult chimpanzees and humans.

COMPARATIVE PSYCHOLOGY

Comparative psychology takes its name from the goal of *comparing the behaviour of different animal species*. The field began shortly after the publication of Charles Darwin's books *The Origin of Species* and *The Descent of Man* Darwin suggested that animal species changed over time, with later generations displaying the characteristics that helped earlier generations survive and reproduce. Arguing that the same process shaped human intelligence, Darwin thought it was logical to trace the development of intelligence in different species, leading up to the development of intelligence in humans. Here is how Darwin put it in his book *The Descent of Man.*George John Romanes (1848-1894), a friend and student of Darwin, wrote a book called *Animal Intelligence* (1882) in which he compared the mental abilities of animals from snails to humans. Romanes used what he called *ejective inference* : the assumption of similarity between animal minds and human minds. Romanes's method of gathering data was to collect *anecdotes* (stories such as would appear in a brief newspaper article or a letter from a friend). Anecdotes are not a very scientific form of evidence, and Romanes knew this, but he felt that if he sifted through the anecdotes for the best and most revealing, it would be helpful in generating hypotheses for later scientific study. This belief that animals had similar mental processes justified Romanes's "ejective inferences." Romanes did not hesitate to *anthropomorphise* or project human qualities upon animals. For example, Romanes would refer to a rat that had just been freed from a cage as happy and carefree. The mental life of a rat was assumed to be like that of a human in the same situation, minus only language. Romanes' attitude is still common. Humans find it natural to project our own experiences into animal minds. Consider the following excerpt from a report in the Science section of the New York Times:

> The similarities between us and Rattus extend far beyond gross anatomy. They're surprisingly self-aware. They laugh when tickled, especially when they're young, and they have ticklish spots; tickle the nape of a rat pup's neck and it will squeal ultrasonically in a soundgram pattern like that of a human giggle. Rats dream as we dream, in epic narratives of navigation and thwarted efforts at escape...

Rats can learn to crave the same drugs that we do — alcohol, cocaine, nicotine, amphetamine — and they, like us, will sometimes indulge themselves to death. They're sociable, curious and love to be touched — nicely, that is. If a rat has been trained to associate a certain sound with a mild shock to its tail, and the bell tolls but the shock doesn't come, the rat will inhale deeply with what can only be called a sigh of relief. Each of the reporter's anthropomorphic claims is based on a research finding. Researchers are usually reluctant to go the last mile and infer such human-like experiences. To the reporter, and most humans, ejective inference comes naturally.

Romanes's reliance upon anecdotes and his tendency to project human qualities upon animals was popular with the public, but it was not endorsed by most American psychologists. At the time, in the 1890s, they were trying hard to make psychology more scientific. They saw Romanes's approach as a step backward toward folk science and speculation.

THE PHYLOGENETIC SCALE

Implicit in the work of comparative psychologists from Darwin's day until about the 1930s is something called the *phylogenetic scale.* This is a ranking of animals according to their general complexity and ability, from lowest to highest. The concept is still present in our language, when we use the phrase "lower animals." The idea that some species are more advanced than others goes back nearly 2400 years to Aristotle's notion of a Great

Scale of Nature or *Scala Naturae*. Aristotle suggested that animals could be ordered from least (for example, worms and snails) to intermediate levels (such as dogs and cats) to the highest and most advanced level (humans). Romanes, like Darwin, accepted this idea as common sense. He wanted to trace the development of intelligence as it moved "up the phylogenetic scale" from simple or primitive animals to complex or advanced animals.

On the surface, the phylogenetic or "phyletic" scale seems reasonable. It is true some species are more advanced than others in particular ways. However, modern psychologists recognise many different forms of intelligence. There could be a different phylogenetic scale for each type of intelligence. On a test of odor recognition, a bloodhound would rate as far more advanced than a human.

The phylogenetic scale makes no sense from an evolutionary perspective. Putting the animals in a linear order (rat, cat, monkey, human) suggests that a rat, if it got a little smarter, would think like a cat. A smart cat might think like a monkey, and a smart monkey might think like a dull human. But why should this be so?

The animals are not ancestral to each other and do not represent a single evolving lineage. No cat ever descended from a rat, no monkey was ever descended from a cat, and no human ever descended from a chimpanzee. All these species have been evolving independently for millions of years. Different currently-existing species do not grow into each other, and there is no reason to expect their mental capacities to fall into a smooth progression.

However, Romanes believed in the *principle of continuity*, which said all animals think the same way and differed only in *speed of learning*. If that was the case, then it made sense to rank animals according to how fast they learned, from least to best, earthworm to human. Today's animal researchers no longer endorse the principle of continuity. Instead, they are respectful of each species' unique forms of intelligence.

MORGAN'S CANON

In 1894, the same year Romanes died, C. Lloyd Morgan published a book with a more sober point of view. Morgan's book was titled *Introduction to Comparative Psychology*. In it, Morgan drew a distinction between:

(1) objectively testable inferences from animal behaviour, which were scientific and

(2) untestable speculations about animal minds, such as Romanes's ejective inferences, which were not scientific.

To use a modern example, you might have a cat that comes running when using an electric can opener. You might infer that

(1) the cat is capable of hearing this sound from the other room and has associated its occurrence with the possible delivery of food. This is scientific speculation because it is based on well-known principles of conditioning, and if you wanted to, you could do experiments to test it, such as associating a new sound to food delivery, or comparing the cat's response to different sounds.

Or you could do ejective inference, and infer that

(2) the cat knows that cans contain food and the can opener is what opens up a can. The ejective inference is not a scientific speculation because it inserts human-level thoughts into the mind of a cat. It also goes way beyond what is necessary to explain the cat's behaviour.

Morgan recommended *economy* or *simplicity* in interpreting animal behaviour. The following declaration from his 1894 textbook came to be known as Morgan's Canon.

In no case may we interpret an action as the outcome of a higher psychical faculty, if it can be interpreted as the outcome of the exercise of one that stands lower in the psychological scale.This idea can be paraphrased as, "Never assume more complexity in an animal's mind than you have to" or "Select the simplest possible explanation for how an animal performs a behaviour."

Morgan's Canon resembles the principle of *parsimony* in science, sometimes called Occam's Razor. Occam's Razor says a simple explanation should be preferred over a complex explanation of a phenomenon, other things being equal.

Anthropomorphisation can be misleading even when it seems to be the simplest explanation for a behaviour. Consider the behaviour of baby rats separated from their mothers and exposed to extreme cold; they emit ultrasonic vocalisations that observers have always assumed were cries of distress "to stimulate the mother to retrieve the pup to the warmth and comfort of the nest." However, this assumption turns out to be false, or only half-true.

Blumberg, Sokoloff, Kirby & Kent found that the cries were actually a means to pump more blood to the heart, when the rat pup is exposed to cold. Using a technique that allowed them to reduce venous pressure, they were able to show that the vocalisations occurred whenever blood pressure dropped below a certain level. The cries resulted in "large pressure pulses in the abdomen" which stimulated warming blood flow and kept the pups alive. The authors noted that their findings "underscore the potential pitfalls of anthropomorphic interpretations". Of course, the cries might serve a dual function, pumping blood and alerting the mother to a baby outside the nest.

Animal researchers of the 1890s were considerably less cautious than modern researchers, when it came to speculating about the "animal mind." Despite his warning against reading too much into animal behaviour, Morgan himself declared that *any form of learning was evidence of consciousness*. A chick might be "unconscious" the first time it made a peck after hatching from an egg, he said, but when the accuracy of the chick's pecking increased, this showed consciousness, "for only by appealing to consciousness can [pecks] be guided".

By 1900 scientists were reacting against such speculations about animal consciousness. Many scientists were ready to stop speculating about "animal minds" altogether. Jacques Loeb argued for a *mechanistic* view. He proposed that animals were like biological machines, and that only humans had minds and consciousness. Loeb explained much animal behaviour as due to *tropisms*, automatic movements toward a stimulus. The movement of moths toward light (or slugs toward beer) would be interpreted by Loeb as a tropism, a simple stimulus-response connection in the animal's nervous system. A stimulus led to a response, he said, and no mind was involved.

Scientists may have been growing cautious in their speculations about the mental abilities of animals, around 1900, but the public showed an undiminished appetite for stories of animal consciousness. Newspapers and magazines published serious articles about dogs, cats, and horses with human-like intelligence. Ten years after Morgan's Canon, many people believed newspaper stories about a German horse that could read and write. The year was 1904 when retired Berlin schoolteacher Wilhelm von Osten presented Clever Hans to the world. The horse could apparently understand German, do arithmetic, interpret calendars, and perform other amasing intellectual feats.

THORNDIKE'S PUZZLE BOX

The American researcher Edward L. Thorndike believed learning occurred through *trial and error*. The animal made many responses, many of them wrong or ineffective, and eventually learned to repeat those that got desirable results. Thorndike was a mechanist, like Loeb. He felt that learning was a matter of creating associations between stimuli and responses, and no speculation about mind was necessary or useful.

To study trial and error learning, Thorndike used a type of test pioneered years earlier by Watson: the puzzle box. Originally, Watson's puzzle box (shown in the illustration) required a monkey to reach through the cage to lift a latch. Thorndike substituted a foot-pedal so the research could be done with cats. If the animal stepped on the switch, the door of the cage opened. The object of the research was to study how quickly the cat learned to perform this response in order to get free. Today this would be called *escape learning*. At first, cats put in the cage explored restlessly, meowed, but did not know how to escape. Eventually they stepped on the foot switch accidentally and the trap door opened. On succeeding trials, they operated the switch faster.

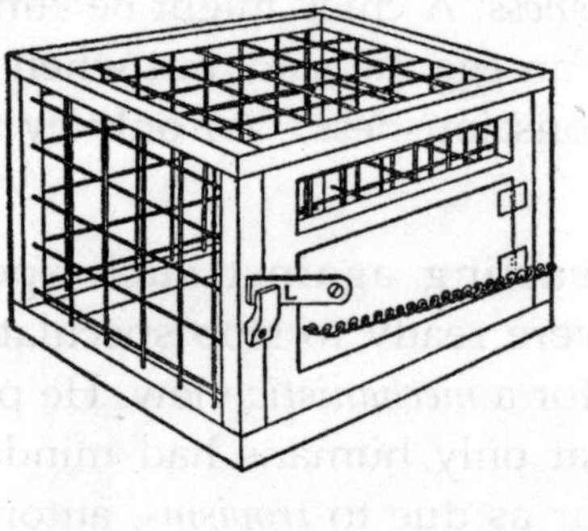

Figure 1. Watson's puzzle box, which required a monkey to reach through the cage to lift a latch

Thorndike explained learning with his *Law of Effect*. Animals tended to repeat a behaviour that resulted in a "pleasing effect." This was an early version of the concept of *positive reinforcement* that B. F. Skinner used so effectively later in the century. Behaviour was varied during a trial and error phase. Thorndike believed that when the animal stumbled upon a behaviour that produced a desirable effect, this created a link or associative bond between a *stimulus* (in this case, being in the cage) and a *response* (stepping on the switch). Later, in the same stimulus situation, that response occurred faster. Thorndike compared the performance of fishes, chickens, cats, dogs, and monkeys in similar tasks. Not all these species could push a foot switch, so he altered the puzzle box as needed for each species, using a response the animal could perform. Thorndike found that different species varied in how fast they learned and where their performances leveled off, but each produced an S-shaped learning curve. This seemed consistent with Romanes' idea that different species learned the same way, but at different speeds.

LEARNING TO LEARN

Harry Harlow proposed a new method for measuring higher learning abilities of animals in 1949. Harlow suggested that humans and other highly intelligent animals not only mastered isolated tasks but also noticed patterns and shortcuts that made them more efficient learners. They not only *learned*, they *learned to learn*, becoming faster at solving new problems as they gained experience solving similar classes of problems.

Learning to learn, which Harlow also called *learning set*, means *picking out a pattern in a series of learning experiences*, so that you *learn even faster* when facing similar situations in the future. An example in humans would be learning how to study correctly for a class. You might find that you do better as you go along, because you have "learned how to learn" in that class.

RAT PSYCHOLOGY

During the 20th Century, American psychologists did a remarkable number of experiments on rats. Given the assumption that all animals learned in similar ways, it seemed logical to focus attention on a species that could be studied easily in the laboratory.

The albino Norway rat or *Rattus norvegicus* had many advantages as a laboratory animal. It was clean, and it was easy to breed and keep in captivity. Specialised breeding laboratories produced rats with a known genetic history. If researchers needed to study litter-mates (rats born from the same parents) laboratories could provide them. The rat was easily tamed, easily handled, and it was smart: a good learner. In many ways it was an ideal research animal for psychologists.

To show you how extreme this trend became, consider that Clark Hull presented his theory of motivationnot just as a theory for psychology, but as a theory for "all social sciences." Yet it was based on experiments with one species: the Norway rat.

Frank Beach criticised the experimental psychologists of his day for focusing on such a small number of species and for putting so much emphasis on the rat. Beach's article in *The American Psychologist*included a cartoon showing a giant rat as a Pied Piper, leading experimental psychologists astray, just as the Pied Piper led rats (and children) away from the city of Hamlin.

R. G. Cook did a survey and found that late 20th Century American comparative psychologists still tended to study three groups: rats, pigeons, and college students. Monkey research had become less common due to a variety of factors including animal rights concerns.

CLASSIC ETHOLOGY

Most of what we have described so far occurred in the United States. A European science of animal behaviour developed independently. It started in the 1870s and a century later merged with comparative psychology to form the modern science of animal behaviour. The European tradition was called *ethology*.

Two names widely identified with the study of ethology in the 1930s through the 1960s were Niko Tinbergen, from the Netherlands, and Konrad Lorenz, an Austrian who worked in Germany. Tinbergen and Lorenz collaborated in the 1930s. Ethology blossomed after World War II when both Lorenz and Tinbergen developed active programmes of research, attracting talented graduate students and co-workers.

Tinbergen and Lorenz invented much of the vocabulary still used today in the field of animal behaviour. They wrote books, published articles, and gave many young scholars a start in the field. For example, Eranus Eibl-Eibesfeldt, the leading proponent of Human Ethology, was a student of Lorenz. Although headquartered in different countries, the two founding fathers of modern ethology collaborated many times during their long careers. Both received Nobel Prizes. They died within months of each other, Tinbergen in 1988, Lorenz in 1989.

In 1951 Niko Tinbergen wrote *The Study of Instinct*, a book filled with entertaining examples of animal behaviour. A year later Tinbergen published an influential *Scientific American* article, "The Curious Behaviour of the Stickleback". For many American psychologists, this was a first exposure to the concepts of ethology.

Tinbergen observed that sticklebacks perform an elaborate mating ritual, carried out the same way each time. First the male stakes out a little area of sand on the bottom of the ditch (or pond, or aquarium). This becomes his territory, defended against all other males. The male stickleback then digs a little hole, shoveling sand with his snout until the hole is about two inches deep and two inches wide. He gathers stringy pieces of algae, coats them with a sticky substance secreted from his kidneys, and piles the algae in the pit, forming a little mound. Finally, the male stickleback wiggles through the mount, leaving a tunnel. Now the stickleback changes color, becoming bluish white on the back and bright red on the underside.

In this colorful, conspicuous dress the male at once begins to court females. The females, in the meantime, have become fat with hundreds of eggs. When the male sights a female, he darts toward her then veers toward the nest. She follows in a distinctive head-up posture. The male leads her up to the nest and thrusts his snout into the tunnel. The female then slides into the nest, her head sticking out one end, her tail out the other. The male prods near the base of her tail with his snout, and this causes her to lay eggs in the nest. When she swims out, the male swims into the nest and fertilises the eggs. He repeats this ritual with several females.

As his mating urge wears off, the male's colors gradually fade and he takes up a sentry position near the nest, fanning the eggs with his fins in order to keep them supplied with oxygen. As the eggs mature, they need more oxygen and the male spends more time fanning them. Finally the babies hatch and the male rounds up strays by catching them in his mouth and carrying them back to the area of the nest. When the babies become large enough, they wander off.

Observations in the Wild

Ethological study starts with an *ethogram*, a simple descriptive list of behaviours, compiled by an observer in the natural environment. Such observations should be carried out in the natural environment because many behaviours make no sense unless they are observed in the context in which they evolved. After observing animals in the natural environment, ethologists may develop hypotheses which are tested in more controlled research, either outdoors or in a laboratory environment. While compiling an ethogram ethologists sometimes build shelters that disguise the observer while allowing observation and picture taking. These are called *blinds*. An ethologist may spend hours in a blind every day, for weeks, simply observing and photographing animals in their natural habitat. The result is a detailed record of natural behaviours. From this one can compile a list of species-characteristic behaviours called an *ethogram*.

A blind is not necessary if animals become *habituated* to human observers. Habituation occurs when humans put themselves in the presence of animals but never do anything alarming. Eventually the animals cease to respond to humans as dangerous. The idea of habituation is not to make animals tame or interact with them as friends; it is to become *irrelevant* so the animal will behave naturally and permit observations that are not distorted by the presence of an observer. Some television programmes such as Meercat Manor depend almost entirely upon habituation.

Jane Goodall was the first chimpanzee observer to use habituation as a technique. She followed a chimp family in the Gombe Reserve, in Tanzania. Most of the her first few months were spent observing the chimps with binoculars, when she could see them at all, from a hilltop where she was in full view of the chimps. Eventually the chimps habituated to her presence and allowed Goodall to approach them for closer observation.

What happened when Goodall interacted with the chimps?Goodall was thrilled when some of the chimps initiated gentle grooming behaviour with her, a sign of friendship among primates. She experimented with feeding bananas to the chimps, a practice that resulted in some interesting observations but also some wild feeding frenzies. Worse, the chimps started invading her camp, looking for food. Having totally lost their fear of humans, they became destructive. Goodall eventually realised she had made a mistake when she allowed her interaction with the chimpanzees to move past habituation to social contact. She instituted new rules of non-contact between human and chimp that are in effect to this day at the Gombe Reserve chimp research facility.

Advancing technology has permitted many ingenious new techniques for observing natural behaviour in the wild. "Creature cams" are small cameras with wireless transmitters that can be attached to anything from a mouse to a whale, giving researchers a video record of a creature's travel. Fiber optics allow filming inside tiny burrows, tunnels, and nests. Global positioning systems permit precise tracking of individual animals, permitting researchers to discover the typical range of endangered species such as the Siberian Tiger (in that case encouraging the Russian government to set aside a huge tract of land as a nature preserve). For some species that are especially shy of human beings, infra-red triggered cameras are the best way to make observations, because human observers can be far from the area when the camera is activated by the body heat of a nearby animal.

Action Patterns (Motor Programmes)

Action patterns or motor programmes are what used to be called instinctive or innate behaviours. They are shaped by the animal's genetic heritage, "wired in" to the nervous system. In many cases these behaviours can be elicited by direct brain stimulation.

A motor programme or action pattern is a *distinctive, stereotyped pattern of movement carried out by most healthy members of a species*. Such behaviours are *species-typical* but not unique to one species. Many motor programmes are shared by a wide variety of species. Consider *yawning*. Humans yawn, hamsters yawn, rabbit's yawn, horses yawn. This is an excellent example of a motor programme: a stereotyped, built-in pattern of behaviour. It is "species-typical" (typical of individual species) but definitely not *unique* to a species.

The phrases *instinctive behaviour, innate behaviour, action patterns, motor programmes* and *wired-in behaviours* have all been used to label species-typical behaviours. *Motor programme* is perhaps the most widely used term now, usually (everywhere but the United States) using the British spelling *motor programme.*

Because motor programmes are genetically based and built into the nervous system, they may persist even when the reason for their existence is gone. For example, declawed cats continue to make claw-sharpening movements that no longer serve a purpose. Well-fed dogs that are raised indoors will "bury" plastic bones in clothes or bedsheets. How does a scientist determine whether a behaviour is learned from other members of the species, or hard-wired into the nervous system? The classic technique is to do an *isolation experiment*. In this type of experiment, a young animal is *raised from babyhood in isolation from other members of its species*. If a species-typical behaviour still emerges under these conditions, then clearly the behaviour was not learned from other members of the species, and the tendency to perform that behaviour is programmed into the nervous system.

Sometimes an isolation experiment occurs accidentally, when a wild animal is reared from infancy in a human household but species-typical behaviours nevertheless appear. When we talked about animal behaviour in class, it reminded me of my roommate's pet squirrel. Cindy has had the squirrel since it was a little baby. He fell out of his nest even before his eyes were open. The squirrel, named Scooter, still shows normal squirrel behaviours even though he has been raised by humans. Cindy made him a little tent by putting two pillows together. Scooter will hide all sorts of toys and food in his house and goes through all the motions of digging a hole to hide his horde in. He loves to run through tunnels and holes and loves to jump from chair to chair. High places are his favorite, e.g. curtain rods and ceiling beams. He rotates his food in his paws before he eats it. Even though he retains these habits, he seems happy in captivity and always runs back to the house after being put outside.

Although the basic form of species-typical motor programmes is built into each animal's nervous system, the motor programmes of birds and mammals are usually modified by experience. For example, bird song can be modified in complex ways, depending upon what a baby bird hears as it grows up. All mammals learn. By contrast, reptiles emerge from eggs with all their motor programmes ready to go. They require no practice.

Releasers

Species-typical behaviours are often triggered by specific stimuli in the environment. The stimulus that triggers an action pattern is called a *sign stimulus* or *releaser*. It is a specific input that activates a motor programme. For example, Tinbergen found that male sticklebacks would attack anything red during mating season, when their own bellies were red. Usually this resulted in attacking other male sticklebacks that were competing for nesting territory. However, the fish would attack anything red, not just other sticklebacks.

In one experiment, Tinbergen fashioned a series of stickleback models, ranging from a very realistic but colorless model to a very unrealistic blob with a red belly. The male stickleback ignored the realistic fish without the red belly but attacked all the models with the red bellies. Evidently the color red released the attack response. For similar reasons, fishing lures with distinctive markings can be very effective. They may release a biting response even if they do not resemble a living animal.

The critical feature of a sign stimulus becomes especially clear when a motor programme is set off by the wrong thing. One day Tinbergen's lab students noticed all the sticklebacks attacking the sides of their aquariums near a window. Across the street was a red postal van. The small patch of red visible through a window was enough to set off attack responses in the Sticklebacks.

Why would the stickleback respond so vigorously to any red object? It is *better to attack all red objects*—wasting a little energy on red leaves—*than ever to fail to attack another male stickleback* during mating season. A failure to attack might allow the second stickleback to slip into the nest and fertilise the eggs. If that happened, the first stickleback would spend its time and energy aiding the reproductive success of another fish. So the fish is biased toward a strategy of attacking *anything* red, resulting in many false alarms but also relatively few failures to drive off competing male sticklebacks.

One lesson from the stickleback research and similar studies by Tinbergen and Lorenz was that *complex motor stimuli are often set off by highly specific stimuli*. Consider the following figure, adapted from Alcock. It contains two examples of complicated movements that normally help birds pass on their genes.

(A) "A male red-winged blackbird copulating with a mount consisting of the tail feather of a female raised in pre-copulatory position." Even though the tail feather is on a seed pod, not a female bird, it sets off the sexual response in the male.

(B) "Willow warblers attacking the stuffed head of a cuckoo." Even though the head is on a stick, clearly not alive and not a real bird, the small warblers nevertheless attack it.

In the proper context each behaviour would seem purposeful or intelligent. But the examples show that these motor responses are triggered by specific stimuli in more or less robotic fashion.

The Greylag Goose

Another example of an action pattern triggered by a sign stimulus appears in so many textbooks it might be called the Pavlov's Dog of ethology. This is the *greylag goose retrieving her egg*. The example comes from Lorenz and Tinbergen.

The sight of an egg outside of the nest sets off egg-retrieval. When a human puts an egg near the nest, the greylag goose sees it but does not immediately reach for it. She does a "double take." She looks away from it and then looks back. Next she moves her head back and forth toward the egg in several short, quick movements. Then she extends her neck as far as it will go-quite far, in a goose-and moves toward the egg with her neck fully extended, using a curious low-to-the-ground waddle unique to this situation.

When the bottom of her beak touches the egg, the goose's neck muscles quiver with tension and slowly she contracts her neck, rolling the egg back toward the nest. The egg, being egg-shaped, tends to roll off to one side or another. The goose moves her beak from side to side, keeping the errant egg on course. If it eludes her, she returns to the nest, spots the egg still lying outside the nest, and repeats the whole movement.

Lorenz found that if he removed the egg while the mother bird reached for it, the bird nevertheless went through the entire action of retrieving the egg, as if retrieving an imaginary egg. This led Lorenz to formulate the concept of *endogenous running-out*. Endogenous means coming from within. Once the action pattern was set off by a sign stimulus, the motor programme kept running to its conclusion, even if the egg had rolled off to the side. If we accept Manning's view that *reflexes* are good examples of motor programmes or action patterns, there are abundant examples of endogenous running-out in human action patterns. Vomiting, for instance, has an all-or-none character. One can hold it off for a while, but once the activity starts, one is forced to carry it through to completion. Many but *not all* action patterns have this characteristic of endogenous running-out.

Supernormal Stimuli

Lorenz once put a football-sized model egg by the nest of a bird. The bird tried to retrieve it using the same action pattern it would use for a normal sized egg. If a normal egg was placed alongside the giant one, the bird made fruitless attempts to retrieve the big egg while neglecting its own normal-sized egg. Lorenz called the exaggerated sign stimulus a *supernormal* stimulus. Supernormal sign stimuli are *bigger* or *more intense* than normal. They elicit a larger-than-normal response from the animal. Why would a bird prefer a grotesquely large egg to a natural one? Manning speculates that larger eggs are usually healthier, and giant eggs never occur in nature, so there is no need for an upper limit on the animal's preference for larger eggs. The animal generally improves its genetic success by retrieving a larger egg first.

Do humans respond to supernormal stimuli? Perhaps. Humans have a biologically "wired-in" response to the human. (Babies at the age of two months will smile at an oval shape with two dark circles where eyes should be.) If the human face is a sign stimulus or releaser, then *cosmetics* create supernormal stimuli. Rouge exaggerates rosy cheeks; lipstick exaggerates red lips; eyebrow pencil and eyeliner exaggerate the darkness of eyes; false eyelashes exaggerate eyelashes. A made-up face may elicit a larger than normal response.

Imprinting

Konrad Lorenz lived on a farm with his beloved greylag geese. He noticed that young goslings *followed the first thing they saw after hatching*. Lorenz called this phenomenon *imprinting*. In normal circumstances, this "works" because the first thing a baby bird sees is the mother goose. However, if the gosling sees a human first, they follow the human as if it were the mother. In a famous picture on the cover of *Life* magazine, Konrad was shown walking on his farm with a string of goslings following him. The *Saturday Evening Post* responded with a picture by the artist Norman Rockwell, showing ducklings following a football. It is true that ducklings will imprint on the first thing that moves, when they hatch, whether it is animate or inanimate. Imprinting is found mostly in *ground dwelling birds* and prey species such as goats and lambs. It keeps the young in close proximity to a protective parent.

Babyishness

Lorenz noticed that baby animals of many species have the same look. Babies typically have large heads in proportion to their bodies. They have large, round eyes, protruding foreheads, soft jaw lines and plump cheeks. Often they are fuzzy, and they have stubby legs. Lorenz labeled this pattern of features *babyishness*. He pointed out that we respond to "cute" features with affection and good humor, whether they are in human babies, babies of other species, stuffed animals, or cartoon characters.

Figure 2. Babyishness in different species

Stephen Jay Gould perceived an evolution toward babyishness in the features of Mickey Mouse. When Disney first drew Mickey for *Steamboat Willy* Mickey's snout was long, his forehead was small, and his behaviour was aggressive. Mickey created music on the steamboat by pulling tails and otherwise torturing various animals, making them cry out. By the 1950s, Mickey's behaviour had mellowed and his appearance had become more babylike. His ears were moved back on his head (creating a larger cranium). His eyes were enlarged by making the former eye into a pupil in the new eye. The snout was shorter, less rat-like. He was given trousers, which made his legs look stubby. In many ways, Mickey became "babyish" and therefore cuter.

Vacuum, Displacement, and Redirected Activities

Action patterns are frequently triggered by a highly specific stimulus, which the ethologists called a sign stimulus or releaser. However, sometimes action patterns appear for no obvious reason at all. *Vacuum activity* is the name Lorenz gave to behaviours set off for *no apparent reason*, "in a vacuum." Lorenz suggested that animals have a need to exercise biologically natural behaviours, even if the behaviour has no function. For example, Lorenz kept a fly-catching bird as an indoor pet. Sometimes he let the bird fly around the room for exercise. He noticed that, although there were no insects present, the bird snapped at imaginary insects in the air. There was no reason to do so; the bird was just exercising its instinctive action pattern. Lorenz called this a vacuum activity. Similarly, squirrels raised from birth in a metal cage will go through the entire sequence of nut-burying activities, despite the lack of dirt or a nut in the cage. The squirrel scratches rapidly on the metal cage floor, digging an imaginary hole, takes its imaginary nut and buries it in the imaginary hole, finally patting the metal floor as if pushing dirt and leaves over a buried nut.

Displacement activities as described by Lorenz are motor programmes that seem to discharge tension or anxiety. For example, if one is trying to entice a squirrel to come up and take a peanut, the squirrel becomes conflicted-caught between two incompatible drives. It wants the nut, but it fears humans. The squirrel is caught between *approach* and *avoidance* tendencies, but it cannot do both at once. It becomes visibly edgy. It may take a few hops toward the human holding the peanut, then scratch itself suddenly or make a few digging movements. This does not mean the squirrel itches or needs to dig a hole. Lorenz suggested it was "breaking the tension" caused by competing urges.

Humans perform displacement activities. One study involved a hidden video camera in a dentist's office waiting room. People waiting to have cavities filled showed all sorts of displacement activities, scratching their heads, stroking non-existent beards, wringing their hands, tugging at earlobes, flipping through magazines at one page per second, and so forth. People waiting for X-rays or teeth cleaning showed fewer of these activities. Like the squirrels approaching a human holding a nut, patients waiting to have cavities filled were caught between two contradictory impulses. They wanted to get the cavities filled, but they probably wanted to leave, also. So they performed nervous activities.

Redirected activity is a third example of action patterns aroused in unusual circumstances. Lorenz defined a redirected activity as a behaviour that is redirected from a threatening or inaccessible target to another target that is more convenient or less threatening. For example, flocks of chickens form a *pecking order*. Chickens form a rigid dominance hierarchy based on status differences respected by all animals in the group. In a chicken coop, each chicken has some other chickens it can peck (because they are less dominant) and some it cannot peck (because they are more dominant, usually larger). At the top of the hierarchy is a chicken that can peck all the others but gets pecked by nobody. At the bottom of the pecking order is a chicken, usually scrawny or unhealthy, which gets pecked by all the others. Sometimes the chicken at the bottom of the pecking order dies from this treatment.

Decline of Old Assumptions

A basic assumption of American comparative psychology in the middle of the 20th Century was that learning involved associations between stimuli and responses, and *all associations should be made with equal ease*. However, this way of looking at animal intelligence was challenged by the ethologists. Ethologists pointed out examples in which learning occurred very rapidly, or not at all, because of the implications of particular acts of learning for survival of the species.

In one experiment, Tinbergen put a ring of pinecones around a digger wasp nest. The wasp spent longer than usual circling the nest when it came out of the hole, then it flew off to find food. When the wasp was gone, Tinbergen moved the pinecones, putting them around a different hole that he made by poking a stick into the ground. When the wasp returned, it went to the new hole that had cones around it. Tinbergen pointed out the implications of this. Apparently the wasp memorised the objects in the vicinity of the nest in just a few seconds, while circling before her flight to find food.

Why would a wasp be so "smart" at memorising objects near the nest? The answer is obvious. The wasp's survival and reproductive success depends upon this skill. The same was true of its ancestors. Digger wasps that memorised the objects near a nest could find their way back to it, reproduce, and pass on their genes (including the genes that controlled this specialised ability). This does not indicate that digger wasps are generally smarter than other insect species, just that they have a special talent adapted to their life circumstances. This is called adaptive intelligence.

Examples like the digger wasp contradict a bedrock assumption of old-time behaviourists, the idea that everything is equally easy to learn, which can be called the *equivalence of associations* assumption. Behavioural psychologists from Pavlov to Watson to Skinner believed that all stimuli were equal, when it came to learning. Learning was assumed to involve links or *associations* between stimuli and responses. Any stimulus was supposed to be as good as any other stimulus, for forming an association.

By the 1970s, psychologists had awakened to the fact that *not all associations between stimuli are equally easy to learn for every species*. If animals are required to make a natural response to a situation, they learn quickly. If forced to make an unnatural response, they learn slowly or not at all. Here is another example of how not all associations are equally easy to learn. Pigeons have a difficult time learning to *peck a key to avoid a shock*. Avoidance learning is usually easy to establish, but not in this case. Pecking is a behaviour associated with *feeding*. It is not natural for a pigeon to peck something to avoid pain. A bird's natural response to a stimulus warning of danger is to fly away. Pigeons learn very quickly, in a single trial, to *fly off a perch* in response to a signal predicting a shock.

On the other hand, pigeons easily learn to peck a key to get food. They even "train themselves" to peck a key for food, a phenomenon called *auto-shaping*. If put in an operant chamber where pecking a lighted key results in grain delivery, pigeons learn to peck the key when it lights up without any further encouragement by humans. Learning which occurs quickly and easily because of how the animal's nervous system is constructed is called sometimes called *prepared* learning.

Instinctive Drift

Not only are natural behaviours easier to learn; they can override the influence the effect of more artificial, trained behaviours. For example, animals can be trained to do cute, human-like behaviours, using operant conditioning techniques. However, over time, the performance may deteriorate as bits of species-typical behaviours intrude on the performance. Breland and Breland called this "instinctive drift."

The Brelands trained animals for roadside tourist attractions. Using standard conditioning techniques, they trained animals to perform complex behaviours for food reinforcement. But in each case, after the behaviour was established, it was disrupted by the intrusion of some instinctive behaviour used by that species to gather or prepare food.

For example, they successfully taught a raccoon to deposit wooden coins into a metal container for food reinforcement. But soon the raccoon started rubbing the coins together and dipping them (not dropping them) into the container. It was performing the motor programme raccoons use to "wash" food in a stream. This interfered with the trick to such an extent the Brelands had to give up on it. Instead, they trained the raccoon to "play basketball." The basketball was so large that the raccoon did not attempt to wash it.

Similarly, a pig was reinforced with food for dropping large wooden disks into a piggy bank. It successfully learned this task. But soon the pig began dropping the coin on the way to the piggy bank, pushing it through the dirt with its nose, and flipping the coin up in the air. This is a species-typical behaviour of pigs called *rooting*. Again, the intrusion of a species-typical behaviour actually prevented the pig from completing its task and receiving food reinforcement.

The Brelands recognised that the phenomenon of instinctive drift contradicted 1950s-era reinforcement theory. The species-typical behaviours appeared even though they *prevented a hungry animal from getting food reinforcement*. That violated the laws of learning carefully built up over half a century. Those laws asserted that behaviours followed by food reinforcement should be strengthened, while behaviours that prevented food reinforcement should be eliminated.

Observations by people like the Brelands helped to sway American psychologists toward the viewpoint of the ethologists. Researchers who used operant conditioning techniques began to explore the relevance of species-typical behaviours and to document the importance of taking the animal's natural tendencies into account. Nowadays, behaviour psychologists who use reinforcement techniques are well aware that some types of behaviour (those prepared by evolution) are easier to train than others.

Phobias as Prepared Learning

Thorndike was one of the few early researchers to recognise that some behaviours were natural, others were not, and therefore some were easier to learn than others. Thorndike pointed out that trial and error learning occurred fastest when animals were:

(1) motivated;

(2) prepared to learn; and

(3) paying attention to the relevant cues.

Thorndike called these "secondary laws of learning." They were ignored in Thorndike's day, then rediscovered about 50 years later when the concept of biologically prepared learning became popular.

Seligman revived Thorndike's *preparedness* concept to explain some strange things about *phobias* (powerful, irrational fears). Seligman asked, Why are some phobias so much more common and difficult to treat than others? He pointed out that the most common phobias involved spiders, snakes, and small animals such as rats. These were also the most difficult to treat. However, in our modern world, more people are hurt by hammers and electrical outlets than by spiders and snakes. Why do psychologists hear no complaints about hammer phobias, or electrical outlet phobias?

Seligman suggested that the common phobias must be biologically prepared by evolution. Fear of snakes, spiders, rodents, heights, water, enclosed spaces...all these things can save a person's life (and preserve the DNA that encourages such a fear response). By contrast, we do not fear hammers and automobiles because they were not a threat to our ancestors, so we have no built-in bias against them.

The Garcia Effect

Taste aversion violates several of the old assumptions about classical conditioning. The gap between the cue (CS) and the biological response (UCR) can be very long, the aversion to a food can be learned in just one trial, and taste aversion violates the assumption of the equivalence of associations, because the illness is almost always blamed on food, even if it is due to some other factor such as a flu virus.

The highly selective nature of food aversion is called the Garcia Effect. John Garcia showed that animals associated illness with food, even if the illness was caused by something else. If rats got sick from a dose of radiation after drinking *saccharin-flavored water* in a cage illuminated with *red light*, the rats later avoided saccharin-flavored water. But they did *not* avoid red light. Similarly, they got shocked after tasting the water, they learned to avoid the environment where they got shocked, but they did *not* learn to avoid the water.

Garcia's claim that the equivalence of associations idea was false seemed like an extraordinary claim, in the 1960s. He ran into skepticism when he tried to publish it in the prestigious journal *Science.* However, Garcia had extraordinarily good evidence. He documented the phenomenon in over 20 studies. He satisfied the skeptics that his findings were not a fluke or the result of methodological errors. Garcia eventually received an award for his work from the American Psychological Association. The Garcia Effect became widely known and is now cited as a prime example of prepared learning or, as it is sometimes called, *biological constraints on learning.*

ARE ANIMALS CONSCIOUS?

Animals commonly look as though they have emotions, intentions, plans, and reactions...things we humans associate with thoughts and consciousness. However, such speculations can lead us astray. How much can we read into animal behaviour, before we start to make mistakes like the comparative psychologists of the 1880s?

There is one thing we can say with assurance: animals do not think in words. This by itself means that cats and dogs and horses do not possess anything quite like human intelligence or reflective self-consciousness. Animal awareness, whatever it is like, must be more immediate and reactive. Humans grasp thoughts and manipulate them like hand tools, using symbol systems. Animals must live more "in the moment," with awareness that is dependent upon immediate stimulation.

It is probably fair to assume that non-human animals have a sort of non-linguistic consciousness. This seems obvious to many pet owners and people well acquainted with species like horses that interact intelligently with humans. Most mammals have limbic systems that rival the size of human limbic systems, in proportion to their brains, and the limbic system is the seat of emotion. With experience, it is easy to get to know and understand the emotions of almost any mammalian species. Sociable birds such as parrots and penguins also seem remarkably transparent in their feelings and motives, to people who know them well. However, humans are notoriously prone to anthropomorphising (projecting human qualitiies into non-humans), and to say that animal awarenes is "obvious" or "transparent" is not to make a scientific argument. Therefore, any sort of objective measure that sheds a light on animal cognition is especially valued by research psychologists.

The Mirror Test

Gallup is famous for arguing that animals, particularly the great apes (chimpanzees and gorillas) do indeed have human-like consciousness. In one famous series of experiments, he studied the ability of apes to recognise mirror images of themselves. He assumed that if they recognised their image in the mirror, they had some kind of self-awareness.. That can be debated, but at least it is an objective measure. First Gallup gave chimpanzees 10 days of exposure to full-length mirrors. During the 10 days, frequency of "other-directed behaviours" such as threat displays directed toward the mirror rapidly fell to near zero. Then he anesthetised the chimps and painted the top half of an eyebrow ridge with "bright red, odorless, non-irritating, alcohol-soluble dye." The marks were located so the animals could not see them without the mirror.

When the chimps awoke from the anesthesia, they were placed in front of the mirror. They increased their frequency of mirror-looking behaviour by over 25 times. They touched the red spot, and they smelled and examined their fingers after touching the red area "even though the dye had long since dried and was indelible". Gallup and other investigators found that a variety of other primates, such as baboons, macaques, mandrills, and two species of gibbons did not respond this way. Gallup and his colleagues concluded that this ability was unique to humans and great apes.

A team of behaviourists, Epstein, Lanza, and Skinner, could not leave this claim of "animal consciousness" unchallenged. What if a pigeon could be taught to do the same thing? Would we attribute self-consciousness to a bird? They decided to use conditioning techniques to see if they could teach pigeons to behave like Gallup's apes.

First the researchers taught pigeons to peck white flecks of paper off their own bodies, using a shaping procedure. Then they taught them to peck at things they could see only in a mirror. Finally they fitted the pigeons with collars so the birds could not see their own bodies except in a mirror. Then they put flecks of paper on their bodies where they could not see them, except in the mirror. Sure enough, the pigeons reached around the collar and pecked off the flecks of paper that they could see only in the mirror. Evidently they referred the mirror image to their own bodies. The researchers triumphantly concluded that Gallup was wrong to use this task as an operational definition of self-awareness, because pigeons would not have self-consciousness of the sort Gallup was claiming for apes.

There is one problem with the Epstein, Lanza, & Skinner rebuttal to Gallup. The researchers seemed to be making the same assumption in their own work that they were criitcising in the work of Gallup: that the "same" behaviour in two different species represented the same subjective events. Pigeons could do the mirror task, and pigeons probably lack self-consciousness, ergo apes who could do the mirror task were not necessarily self-conscious. This logic only works if pigeons and apes go through the same cognitive processes in performing the task, and we do not know that.

Gallup's experiment has been criticised on various other grounds. Westergaard and Hopkins noted, "Only 20%...of the chimpanzees he actually tested actually passed the notorious mark test." They also pointed out that "effects of human contact on the development of self-recognition in chimpanzees and orangutans have yet to be assessed." In other words, this may not be a spontaneous or natural behaviour. Finally, they point out that monkeys, too, readily learn to use mirrors to guide their own hand movements to obtain hidden food, so perhaps the discontinuity between monkey and apes (emphasised by Gallup) is not as great as first appears. Despite the criticisms, Gallup's research has definitely been successful in one way: raising questions, stimulating thought, and inspiring new research.

Animal Information Processing

Comparative psychologists have been very creative in recent decades. Many of their experiments use conditioning techniques to show that animals perform sophisticated information processing, with results similar to humans. How have comparative psychologists been creative in recent decades? What are the two alternatives, when animals are trained to do "clever" things? Is this the return of the "similarity assumption" of Romanes? Not necessarily. There are always two possibilities when non-human animals perform in a way resembling humans:

1. The animal is carrying out the same information processing as humans.
2. The animal is performing the same task in a different way.

Either alternative is interesting to psychologists. The first possibility, that a common mechanism underlies the behaviour of different species, might point us toward fundamental or "primitive" processes widely shared in the animal kingdom. Complex cognitive activity is composed of many simpler processes, and surely humans and different species share some of those.

The second possibility, that different species (or different individuals) perform a task in different ways, also presents an interesting challenge. Scientists can compare the different solutions for clues to their advantages and disadvantages. For example, one solution might involve only a few neurons; therefore it might be very efficient. However, the efficient solution may be limited in other ways.

Here are some examples of comparative research showing that different species can perform some of the same information processing tasks as humans, whether or not they do it in the same exact way.

— Kuhl and Miller showed that a chinchilla was just as good as humans at discriminating complex human speech sounds...an ability once thought to be a unique evolutionary adaptation of humans.

— Brannon and Terrace showed that chimps could learn to count from 1 to 9 and to identify pictures that contained 1 to 9 different objects.

- Blough showed that pigeons saw letters of the alphabet much like humans. The pigeons produced a "confusion matrix" (tendency to mistake one letter for another) similar to humans.
- Geiger and Paggio showed that common houseflies are fooled by the Muller-Lyer illusion, perceiving the midpart of the bottom arrow as longer than the midpart of the top arrow.
- Bottle-nosed dolphins were shown to have a memory for lists of sounds similar to that of humans.
- Goldfish were shown to see the "same" color even when spectral composition is radically altered, just like humans.

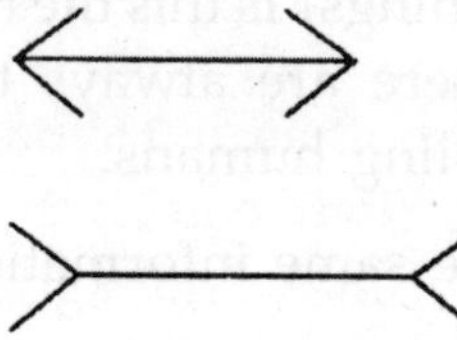

Figure 3. The Muller-Lyer illusion

- Tree swallows follow a logical "game-playing" or "decision-making" pattern called Tit-for-Tat.
- Cats have the same P300 component of the EEG, equated with meaningful interpretation of stimuli, as do humans. So do monkeys.
- Octopi are capable of observational learning. If allowed to watch another octopus that has been trained to pick one of two targets, the observer will pick the same target.

And the list goes on. Many laboratory tests of human cognitive abilities can be adapted to non-human animals, with suitable ingenuity. The result is often the discovery that non-human animals perform similarly to humans.

Adaptive Intelligence

Animals will generally be "smart" in ways that are adaptive for the species. If a certain form of information processing makes members of a species more likely to reproduce and pass on their genes, that type of intelligence will be positively selected in that species. In what ways will animals be "smart," generally? Why are pigeons good candidates for a visual discrimination test? Why are rats so good with mazes?

For example, in explaining why pigeons perform letter discrimination tasks as well as humans, one must remember that pigeons are "visual animals" because birds must spot things from the air to survive. A pigeon is a good candidate to make sophisticated visual recognition judgments in the laboratory, because it must do so in nature. A rat would probably be a poor subject in visual recognition tests, because a rat depends more on taste, touch, smell, and spatial ability.

Rats have dexterous little paws. They are "smart" in using them to manipulate objects. Pigeons would do not well in tests that involve handling objects. Rats are great at running mazes. Why? This laboratory task resembles their activity in natural environments. Rats make their living by exploring territories and finding food. A pigeon might do poorly in a maze, but it could range over a greater territory than a rat without getting lost. Intelligence is clearly specialised and relates to the adaptive needs and abilities of a species.

Specialised Intelligence of Bees

So-called "lower" organisms often show sophisticated learning abilities, when this aids their survival and reproduction. Bees and their relatives are especially good at learning the location and appearance of flowers.

Bees communicate location information to their fellows with the famous waggle dance first described by Von Frisch. A bee returning to its hive, if it has found flowers, makes a distinctive wiggle of the abdomen and dances up the side of the honeycomb. The bee's angle from the vertical indicates the angle of the flower away from the sun. A dance 100 degrees to the left of the 12 o'clock position indicates flowers can be found 100 degrees to the left of the sun, as seen from the hive.

If the food is found by flying away from the sun, the bee dances downward rather than upward. The duration of the dance tells other bees the distance of the pollen source (a fast dance indicates food close by). Von Frisch also identified other factors that aid bee navigation, such as odor particles borne by the foraging bee that returns to the hive.

Von Frisch won a Nobel Prize for his work, but many scientists challenged it in various ways. Some suggested that bees responded to sounds rather than the dance, although other researchers believed bees were deaf. Some suggested that bees used odor alone to direct their hive mates, and the waggle dance was irrelevant.

These controversies were seemingly put to rest by detailed research carried out by the team of Wolfgang H. Kirchner, of the University of Wurzburg in Germany, and William F. Towne, of the Kutztown University in Pennsylvania. They devised robot bees that delivered messages using waggle dances, or by delivering samples of pollen, or by producing a vibrating sound similar to the beating of wings. Among other things, Kirchner and Towne discovered the following:

How did researchers test the "waggle dance" theory with robot bees? What other discoveries did they make?

— Bees can indeed hear, "and their ears are well suited for detecting the sounds associated with the dances."
— "Both sound and dance are needed to communicate information about the location and the food." A bee with clipped wings, which raised the frequency of the sound produced by the dancer, could not recruit other bees.
— Bees in the audience of the hive may rub their thoraxes against the comb, producing a squeak that vibrates the comb. This causes the dancing bee to stop the dance and dole out samples of food, "so that her audience knows not only the direction and distance to the feeding site but how the food smells and tastes as well."
— A robot bee, dabbed with a faint floral fragrance and "waggled" by a step motor, successfully guides bees to a sugar solution placed in a distant field.

Bees form detailed cognitive "maps." They optimise their routes to flower locations, taking the shortest route when visiting multiple sites. Bees refuse to respond to a waggle dance that points to the middle of a lake. However, they respond to a waggle dance that points to the opposite shore of a lake.

Does it sound unrealistic to suggest that "pictures" and "maps" could be preserved in the tiny heads of insects? The bee brain is tiny compared to a human brain, but it is nevertheless a complex system with over 200,000 different nerve cells. No wonder they can carry out some complex information processing.

Chimp Cognition

Chimpanzees and humans are "more closely related to each other than either is to any other living primate". Perhaps you have heard that humans and chimps are "99% identical" in their genetic structure. That is true, but it is a bit misleading. As Plomin and Kuse point out, the average stretch of human DNA is almost 99% identical to the corresponding stretch of chimp DNA, but small differences in DNA can lead to numerous differences in the proteins generated by the DNA. Comparing proteins from chimps and humans, only about 75% are identical. Still, that is a lot of overlap, so it is not surprising that chimps come the closest to human intelligence of all non-human animal species.

Kohler's chimpanzee Sultan showed "insight" in solving a problem. First he assembled a stick, then piled up boxes, then used the stick to reach a banana dangling from a tree. For a long time, psychologists have been impressed by the intelligence of chimpanzees. Perhaps the most famous example comes from Wolfgang Kohler in his book The Mentality of Apes. Kohler suspended bananas out of the reach of chimpanzees.

He found that chimps could assemble two sticks together to make a single instrument to reach the bananas. They also proved capable of piling up boxes to reach the bananas. And they could combine these techniques when necessary.

Kohler claimed the apes were showing insight, coming up with the idea on their own in a flash of intuition. Epstein, one of the behavioural psychologists who showed pigeons how to use mirror images of their bodies, decided to test a behavioural explanation. If Epstein could teach a pigeon to do something similar to Kohler's apes, maybe people would realise there was nothing mystical or mysterious about it. Epstein could not completely imitate Kohler's observations with pigeons, because a pigeon cannot lift a box. So Epstein had the pigeon retrieve a box from another compartment, climb on it, and peck a small plastic "banana" in order to get grain reinforcement. The experiment worked. First Epstein taught components of the goal behaviour. He taught the pigeon to peck at the plastic banana for grain reinforcement. Then he taught the pigeon to push the box around. Finally, as predicted, the pigeon had the "insight" to put these components together when the banana was hoisted out of reach. The pigeon retrieved the box, climbed on it, and pecked the banana.

Kohler's apes were merely combining previously familiar actions. However, creativity always involves combining pre-existing components into new combinations. Kohler claimed that his chimps came up with the solution on his own, based on past experience with sticks, climbing, and bananas. Epstein's pigeon needed long and careful training. So the differences between the two performances may be just as important as the similarities...it depends what point you want to make. Epstein showed that no mysterious, magical abilities were required to explain insight, and that was Epstein's main goal, so he succeeded. Kohler showed that apes were capable of coming up with creative solutions to problems without special training, which was significant in itself.

Menzel studied spatial memory organisation in chimpanzees by letting the chimp observe food being hidden in 18 randomly chosen places within a yard. Released from its cage, the chimp quickly gathered the food, following an "optimum" route that minimised the distance traveled. This indicated that it was not merely imitating the route that the humans had taken; it had formed a "mental map" of where the food was hidden and was following its own map.

Sands and Wright demonstrated that a rhesus monkey could memorise a list of items. Like humans doing such a task, it showed primacy and recency effects, meaning it showed superior recall of early and late items in the list. The researchers also found that picture memory in monkeys is almost identical to that of humans. Both chimps and humans are exceptionally good at recognising pictures they have seen the day before.

Woodruff and colleagues showed that Sarah, an adult chimpanzee, could pass some of the tests developed for human children by the Swiss psychologist Jean Piaget. Piaget is famous for his demonstration that small children cannot keep track of concepts like liquid quantity. If you fill a narrow beaker full of colored water, a young child will assume it has "more" in it than the beaker next to it which is twice as fat but has a slightly lower level of liquid. Around the age of 5, children become able to pass Piaget's test. They come to realise that the amount of liquid does not change just because it is in a differently shaped container. Sarah the chimp could pass this type of test, too.

Ape Language

Robert Yerkes, a pioneering comparative psychologist who studied primates, wrote about the communicative sounds of chimpanzees in 1925. Yerkes might have been surprised to learn that 50 years later chimps would be pressing plastic buttons on a keyboard, using a language called "Yerkish" at the Yerkes Primate Research Center in Atlanta. Yerkes was right that chimpanzees do better with motor movements than with sounds. The earliest attempts to teach chimpanzees language concentrated on spoken words. Results were discouraging. As Yerkes reported in 1925, chimps have "no gift for the use of sounds."

In the 1930s Winthrop and Luella Kellogg raised a female chimp named Gua along with their own baby. Gua learned to respond appropriately to 100 words but never learned to speak. However, in their own defense, the Kelloggs later pointed out they were not trying to teach her to speak. They were more interested in the experiment of raising a chimpanzee alongside a human baby and comparing their development.

In the 1940s Keith and Cathy Hayes also raised a chimpanzee from infancy. They succeeded in teaching the chimp to mouth some words by using his hand to push his lips into unfamiliar positions. The words that the chimp "spoke" were Mama, Papa, and cup, all spoken in a breathy whisper barely recognizable as speech. The chimp mouth is simply not designed for human speech.

The solution, as Yerkes suggested, was to use signs. A chimp named Washoe was raised from one year of age by Allen and Beatrice Gardner, with the express purpose of teaching her sign language. The Gardners used only American Sign Language around Washoe. Experienced signers visited Washoe, played with her, and signed to her as they would with a pre-school child. The Gardners recorded many controlled experiments on film, providing convincing evidence that Washoe knew the meaning of many signs.

Washoe cussed with the sign "dirty," which was also her name for excrement. She learned to ask for favorite toys (e.g. a doll). She asked to go outside when she wanted to. She asked for tickles and hugs, and she labeled environmental objects. Once she even made up a new phrase. Witnessing a duck for the first time, she signed "water" and "bird" in quick succession. "Water bird" became a famous example of Washoe's creativity in language. At the climax of their research, the Gardners participated in a film titled The First Signs of Washoe that triumphantly heralded the demise of the "language is unique to humans" theory. The film opened with quotations from famous linguists like Noam Chomsky saying (in effect) "only humans talk." Then the movie showed Washoe labeling various objects and situations with correct signs. It was a powerful and effective demonstration. However, another scientist working with chimp language—Herbert Terrace—did not see "talking" in the Washoe film. Analysing the sequences of human/chimp signing in slow motion, Terrace found that Washoe was often imitating her human trainers.

Terrace and a group of assistants decided to repeat the chimp sign language experiment with better controls. They raised a chimp named Nim Chimpsky. The name is a humorous reference to linguist Noam Chomsky. Nim was taught sign language using intensive training techniques similar to those employed by the Gardners with Washoe. Like Washoe, Nim succeeded in learning to use sign language for labeling objects and actions. However, also like Washoe, Nim never mastered the grammar of sentence construction. If Nim wanted grapes, he might emit the sequence of signs "Nim eat grape grape grape eat eat Nim eat." Terrace concluded that chimps can learn labeling but they cannot learn to construct sentences with anything like the competence of a human two year old. Researchers at the Language Research Center, associated with the Yerkes Primate Research Center in Atlanta, refused to give up. Sue Savage-Rumbaugh carried out a long and careful study of language ability in the bonobo. The star student was Kanji. Kanji learned many signs and also showed a definite ability to generate two-word sentences in which order conveyed information, a basic requirement of grammatical speech.

Boysen's Chimps

Psychologist Sarah Boysen of Ohio State University explored the ability of chimps to form concepts such as "more than" and "less than." She found that two of her chimp subjects, Sarah and Sheba, were both capable of learning such discriminations easily. Boysen then tried a variation of the experiments, using gumdrops as stimuli. Gumdrops were some of the chimps' favorite treats. The chimpanzee was presented with two plates of gumdrops. One had more on it, the other had less. For example, one might have five gumdrops, the other three. While the other chimp watched, the chimp being tested was asked to point to one of the plates. Whichever plate it pointed to was given to the other chimpanzee. In this situation, the chimp doing the pointing should have learned to point to the plate with fewer gumdrops on it, in order to have that plate given to the other chimpanzee and in order to get the plate with more gumdrops for itself. Instead, something odd happened. The chimps insisted on pointing to the plate that had more gumdrops, even though this meant that more gumdrops went to the other chimpanzee. Boysen says they seemed to know they were making a mistake when they pointed to the plate with more gumdrops, but they could not stop themselves. Often they expressed frustration immediately after pointing at it, even before Boysen removed the plate and gave it to the other chimp. Boysen herself was surprised that the chimps could not learn to maximise their gain. She said, "It was the first task in 20 years that I'd failed to teach a chimpanzee". Then she tried a simple variation. She replaced the gumdrops with plastic poker chips. Now the chimps had no trouble with the task. They pointed to the plate with fewer poker chips on it. This meant the plate with fewer gumdrops went to the other chimp, and the chimp that did the pointing got the larger number of gumdrops. Before the poker chips were substituted for the gumdrops, the chimpanzees seemed to be at the mercy of their desire for food. "The chimps understood the rule," Boysen says, "but they couldn't act on it" because of some biological imperative to get more food. Moving into the symbolic realm, by using poker chips in place of the gumdrops, allowed them to transcend that biological imperative and use abstractions like "more" and "less" to maximise their gain.

Emotions in Chimpanzees

Language is not everything. Robert Yerkes, for one, preferred to emphasise that chimpanzees had human-like emotions. Yerkes was a gentle man who loved animals. In his book Almost Human Yerkes cites many examples of human-like behaviour in great apes, including chimpanzees. For example:

> Chimpita...took safe refuge in a mango-tree and refused to come to his keeper. "So," says Madam Abreu, "I went to the tree and, speaking to him, pretended that I was injured in the arm and suffering. Immediately, on seeing that I was in trouble, he jumped from the tree, and coming to me held my arm and kissed it strongly. And so we were able to catch him."

Of course, this is an anecdote. It would have impressed Romanes back in 1884. But anecdotes are not very good forms of scientific evidence. They are essentially stories, and who is to say if they are reported accurately? All sorts of outlandish stories exist in the culture around us; we do not accept them as scientific evidence. Why should stories of animal behaviour be any different? Fentress made a different case, arguing that anecdotes are a valuable source of insights into animal behaviour. Faulkes strongly agreed, calling anecdotes a "gold mine" for inspiring future research.

REFERENCES

Grier JW & T Burk, 1992. *Biology of Animal Behaviour*. Mosby Year Book, St. Louis.

Grusec, Joan E.; Hastings, Paul D. *Handbook of Socialisation: Theory and Research*. 2007, Guilford Press.

Holt, John. 1983. *How Children Learn*. UK: Penguin Books.

Paivio, A. 1971. *Imagery and verbal processes*. New York: Holt, Rinehart, and Winston.

Sandman, Wadhwa, Hetrick, Porto & Peeke. 1997. "Human fetal heart rate dishabituation between thirty and thirty-two weeks gestation". *Child Development*, 68, 1031-1040.

6

Animal Learning and Development

Broadly speaking, a *stimulus* is any detectable change in an animal's environment. A *response* is any behaviour or physiological event. Animals have innate or instinctive responses to stimuli. Examples include the way newborn mammals move under the influence of tactile and olfactory signals to find nipples and the way squirrels bury their food when periods of daylight shorten.

The usual technical definition of *learning* or *conditioning*, as it is more correctly described, is any relatively permanent change in response that occurs as a result of experience. Interestingly, this refers to a response and not a cognitive outcome such as knowledge.

Not all changes in behaviour are a consequence of learning. The reference to a *'relatively permanent change'* is added to exclude modifications of behaviour due to motivational factors, physiological variables or fatigue. A thirsty horse that drinks despite having refused water five hours earlier has changed its behaviour but is not considered to have learnt anything in the interim. Instead its motivation to drink has changed as a result of shifts in variables such as blood volume and the concentration of sodium in body fluids. Meanwhile fatigue can change behaviour, transforming a playful kitten into a snoozing ball of fluff but its effects could not be described as relatively permanent.

Because the definition of learning has experience as a prerequisite, it excludes permanent changes in behaviour resulting from maturation or debility. So, when male puppies progress from squatting to leg cocking, they have not learned that this new posture elevates the smelly signal they leave for others but are simply maturing and responding to increased levels of circulating testosterone. The aged stag whose roar is ever weaker during the rut has not learned that the hinds are unimpressed. The old muscles in his rib cage and belly have just given up the ghost.

Instead of relying solely on invariant behaviour patterns for survival, animals living in constantly changing environments thrive if they are able to respond to change. Learning allows animals to use information about the world to tailor their responses to environmental change. By avoiding pain and discomfort, animals can make their life more pleasant. Invertebrates such as flies, slugs and ants show advanced forms of learning when avoiding stimuli that have elicited pain responses. Broiler chickens prefer to consume food that contains analgesics presumably because it ameliorates the subjective state of pain caused by chronic leg weakness. Similarly baboons have been trained to self-inject psychotropic drugs which, it is presumed, improve their quality of life.

The original rules of what we call learning theory first came from the laboratories of psychologists and behaviourists who used clinically controlled, some would say sterile, stimuli. These days the study of animal learning is increasingly the pursuit of cognitive ethologists. These are the behavioural scientists who, when considering the way in which a species processes information, emphasise the importance of the environment for which a species evolved and determine how the biology of a species can influence its behaviour.

Learning is the modification of behaviour in response to specific experiences. Learned behaviours of animals can be classified in various ways:

In Associative Learning an animal learns to associate one stimulus with another. There are 2 forms of associative learning. The first is described as classical conditioning and was demonstrated by Pavlov in his famous dog experiments. First he stimulated dogs to salivate by rubbing meat powder on their lips. The meat powder odor stimulated salivation as a physiological process. He then "conditioned " dogs by ringing a bell, or a tuning fork, at the same time as applying the meat powder. He then demonstrated that the animals had become conditioned to associate the sound of the bell with the meat powder and would salivate to the noise without food present. A second form of associative learning is called operant conditioning. In operant conditioning an animal conducts a chance action e.g. pressing a lever and is rewarded with food. Rapidly the animal learns that the action leads to a food reward and will carry out the behaviour repeatedly for food. This is the type of approach applied in most animal training.

Observational learning or modelling is when the animal learns a behaviour through watching other animals conduct the behaviour. For example,in a pack animal such as the wolf, hunting behaviours, fit this category.

Insight learning is in a sense the "highest form" of learning observed. It is the ability to problem solve or to perform a correct or appropriate behaviour the first time the animal is exposed to a situation. For example a chimpanzee may stack boxes to obtain a food object hung out of its reach without ever having seen this solution to the problem before. However, it is not restricted to primates e.g. Ravens and other birds will also show insight learning.

An animal may cease to carry out a response to a stimulus if the appropariet response no longer occurs. For example using the operant conditioning example given above if pressing the lever no longer leads to a food reward the behaviour will probably become less frequent and stop.

NATURE VERSUS NURTURE

A debate among human psychologists, as well as animal behaviourists, concerns the relative importance of instinctive and learned behaviours. The factor influencing learned behaviours is the environment in which the animal is placed. This debate is often referred to as Nature (genes) versus nurture (environment).

The balance between fixed and learned behaviours varies with species. In humans, a large part of our behaviours are learned. In the absence of role models to learn from (deprivation) there is a greater emphasis on programmed behaviours. Some behaviours need to be fixed for survival because there is unlikely to be a second chance to learn them. For example the kangaroo rat instinctively reacts to the sound of a rattlesnake by executing an escape jump. This is a species specific defense response.

With learned behaviours there ar often critical or sensitive periods for the development of the appropriate learned behaviour. For example a dog that has not been socialised to humans by 14 weeks of age is unlikely to be a good pet. Dogs that have not been well socialised to other dogs may be frightened of them or will not breed with them.

Imprinting is another example of a process that must occur within a distinct, usually short, time period. It is also irreversible and involves an attachment to an object that will evoke subsequentadult behaviours and can be generalised to all examples of the object. The classic example is the work of Lorenz with goslings in which they imprinted on Lorenz. The imprinted adult geese directed courtship behaviour to Lorenz rather than other geese.

LEARNING AND MEMORY

Learning is characterised by persistent and measurable changes in behaviour which are not associated with fatigue, altered motivation, or maturation. Some information or knowledge is acquired and is then used to alter the individuals actions and responses. Learning as an adaptive behaviour allows individuals to adapt to specific environment challenges.

Habituation and Sensitisation

Habituation refers to a gradual decrease in behavioural responses with repeated encounters of a particular stimulus, which proves of no consequence. It depends on a change in the synapse between the sensory and motor neuron. Indepedent of conscious motivation or awareness it aids in distinguishing novel and meaningful information from the general background noise. Sensory Adaptation refers to a change in the responsiveness of a sensory system when confronted with a constant stimulus.

In contrast, sensitisation refers to an increase in behavioural responses following repeated applications of a particular stimulus. Following sensitisation, very little stimulation is then required to produce exceedingly large effects. Initial. light stimulation of peripheral skin receptors may not activate nociceptive (i.e., pain) neurons of the spinal cord.

Continued, repeated stimulation of the same pathways will bring about central sensitisation suggesting the presence of an iritating or potential more damaging skin issue. Tissue damage or continued inflammation may thereby cause chronic pain conditions. In long-term potentiation (LTP), physiological effects of subsequent synaptic signals are strengthened following initial activation. This process is thought to form an integral component of memory and learning. Repeated stimulation of different neural centers can also stengthen coordinated neuronal firing and thereby lead to seizures.

Associative Learning

The ability to learn is a basic foundation of intelligence. Humans are particularly keen observers of the world around them and are thereby able to reliably identify the presence of different regularities and generalisations. Patterns that are of consequence to our daily lives are readily learned, which suggests the presence of powerful biases in their search and representation. We tend to remember information more easily if we have uncovered it ourselves than if they are simply presented to us.

An effort to form mental images out of any material to be learned may often enhance retention. Dual Code Theory suggests that most information can be committed to memory in either verbal/linguistic code or mental image code. The use of multiple codes for harnessing any material thus may enhance performance by increasing the number of retrieval paths.

Classical Conditioning

Classical conditioning, a form of associative learning. It requires an unconditional reflex, where an unconditional stimulus (US) brings about an automatic, unlearned (unconditional) response (UR). If a neutral stimulus (NS) tends to precede it, an association is made and the conditional response (CR) becomes transferred onto the (previously neutral) conditional stimulus (CS); a conditional reflex has been learned. For instance, food (US) elicits salivation (UR) in a dog as a natural response. If the sound of a bell (NS) frequently occurs before the food (US) is presented then the mere sound of the bell (CS) will elicit salivation (CR). A mistranslation of "conditional" as "conditioned" meant that in English the CS and CR were referred to as conditioned stimulus and conditioned response, and the verb "to condition" was derived to refer to the process responsible for the establishment of new CR.

Fear conditioning allows organisms to acquire affective responses, such as fear, in situations where a particular context or stimulus is predictably elicits fear via an aversive context (e.g., a shock, loud noise, or unpleasant odor).

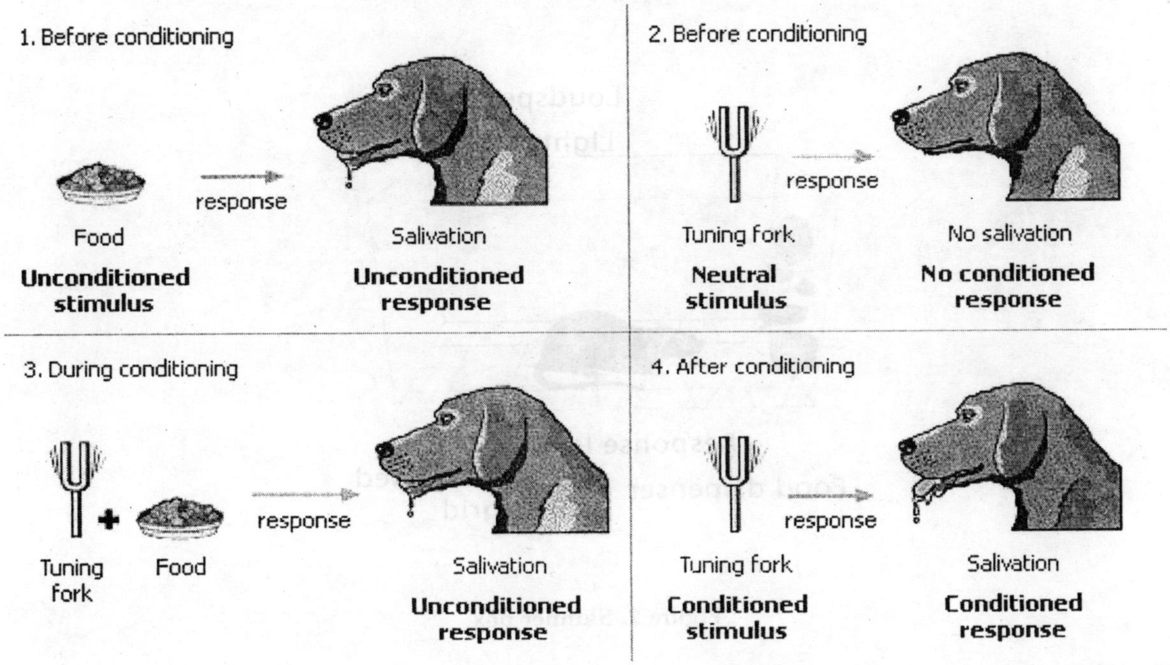

Figure 1. Classical conditioning

Conditioned Taste Aversion (i.e., Garcia conditioning) occurs when a subject experiences symptoms of a toxic, spoiled, or poisonous substance such as nausea, sickness, or vomiting after consuming unfamiliar food. Learned aversion of such taste is an adaptive trait that trains the body to stay away from items that are likely poisonous (e.g., berries or mushroom). This association sometimes occurs in subjects even when sickness was merely coincidental and not related to the food.

Operant Conditioning

Operant conditioning, sometimes called instrumental conditioning or instrumental learning, was first extensively studied by Edward L. Thorndike (1874-1949). Thorndike's most famous work investigated the behaviour of cats trying to escape from various home-made puzzle boxes. When first constrained in the boxes the cats took a long time to escape from each. With experience however, ineffective responses occurred less frequently and successful responses occurred more quickly enabling the cats to escape in less and less time over successive trials.

B.F. Skinner (1904-1990) extended the theories proposed by Thorndike about 40 years after Thorndike published his works. His ideas of animals operating on the environment, lead him to analyse how behaviour is changed by its consequences. The contraption that became known has the Skinner box measured if a task was completed and the duration of time it took for the task to be completed.

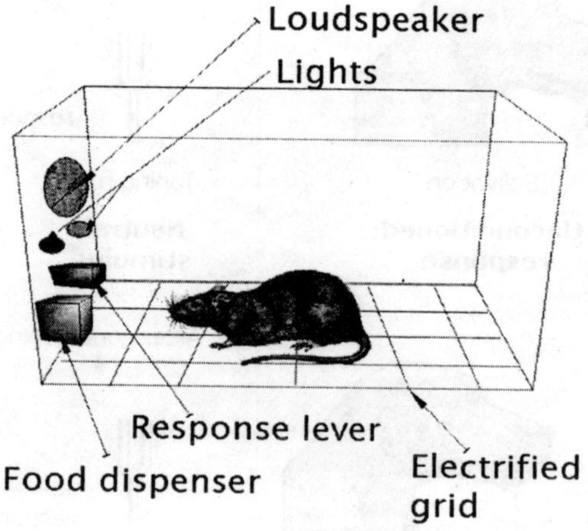

Figure 2. Skinner box

The Skinner box consisted of a bar on the wall that when pressed triggered the release of a food pellet. It was Skinners belief that by rewarding an animal when an appropriate action occurred it would increase the likelihood that the behaviour would be repeated.When the rats, placed in the box, accidentally tapped on the bar on the wall a pellet was released and Skinner observed the amount of time it took the rat to find the pellet. As the rat learned that a pellet released each time he happened to step on the bar, the rat learned to press the bar and to immediately find the food. This training is what became known as operant conditioning (Alcock).

Reinforcement always strengthens the behaviour that preceded it and may consist of any outcome that the animal considers desirable. Reinforcers can broadly be categorised into positive reinforcers (i.e., obtaining access to something the animal wants) or negative reinforcers (i.e., removing something considered unpleasant or painful). Punishment is always designed to discourage further performance of the behaviour it is paired with. Positive punishment achieves that by inflicting an unwanted outcome such as unpleasant sounds or odors, electric shocks, or pain, while negative punishment removes access to something the animal wants. Skinners operant conditioning is seen in the learning behaviour of many animals as well as humans. Often times it goes completely unnoticed, but it can also be used as a deliberate tool in training. As young children explore their world and learn how to eliminate unpleasant stimuli and satisfy desires through proper behaviour they are learning through operant conditioning.

Operant conditioning is used extensively in training animals towards performing quite complex tasks. Shaping involves the step-wise reinforcement of successive approximations. This training method relies on the animal to exhibit an interest in showing a wide range of behaviours as well as in obtaining rewards. When a desired behaviour is performed, the trainer rewards this action. As this method relies on compliance from the animal, most suited to this training are a combination of positive reinforcers (e.g., a small item of food) for behaviour compatible with the trainer's expectations, and negative punishment (e.g., withholding that food item) for behaviour that is not. Initially rewards are issued for even a crude approximation or a component of the behaviour. While these attempts may have earned the animal a reward at the beginning, subsequent rewards are contingent on progressively narrowing the behaviour to what the trainer aims for.

Latent Learning

Latent learning refers to an individual's ability to learn associations without explicit reinforcement. Exploratory behaviour serves to acquire an understanding of the spatial relationships of objects. It includes the formation of "cognitive maps" of the surrounding. Daily success depends on knowldge of spatial relationships. Hummingbirds and bees recall the location and status of harvesting of flower resources, and several species of birds are able to track a large number of seed stores.

Observational Learning

Bandura illustrated the power of an instructional strategy in which children are allowed to observe behaviour of individuals who are significant to them. Subsequently they are likely to imitate their role models especially when such behaviour is reinforced. His work illustrates that observers readily adapt their behaviour based on what they see around them, even when they received no encouragement or incentives to do so. Integrating a continuous interaction between behaviours, cognitions, and the environment, his social learning theory stresses the importance of observational learning, imitation, and modeling.

Language Learning

Human Language

Humans show exceptional skill to communicate with fellow conspecific using verbal and gestural symbols. Such competence is arguably one of mankind's greatest assets as well as a key to most of our species' achievements. It brings us together with our peers and enriches us through an exchange of experiences, thoughts and value systems. It endows us with the means to acquire skills for situations that we may have never personally encountered before. A staggering range of regionally distinct languages and dialects, grouped in larger language families, not only serves as a system of communcation but also tags us for membership within a specific group. In addition to bringing us together, it has thereby been at the heart of an ancient source of division. Although few things in biology ever group cleanly into one of the nature vs. nurture extremes, this particular division seems to be purely cultural. Regardless of the specific race, ethnic or regional group that we happen to have joined through birth, we all are able to acquire competence in any human language. In particular, we seem to acquire our native language without formal education, long before many of our intellectual capabilities have mnatured, and simply by immersion into the particular language environment of parents and relatives during the first few years of our life. The scientific study of the nature and structure of languages is called linguistics.

Universal Grammar. BF Skinner suggested that infants learn language through a process described as operant conditioning, namely, via the monitoring and management of reward contingencies. Skinner's position would be that a four-term contingency analysis comprised of motivating operations, discriminative stimuli, responses and reinforcing stimuli would be the means by which behaviour could be explained. In children and infants this process would be expanded by what he called "shaping", "prompting" and other stimuli modeling, imitation and reinforcing procedures. Language acquisition then is a process that would take thousands of instances of such training, and this appears largely to be what takes place. Critics who do not understand the inductive power of this approach largely assert that this view now "appears quite simplistic." However, this argument to complexity is not dissimilar to arguments against Darwin's theory of natural selection. How could the amasing complexity of the animal kingdom come about through such a simple means? Surely, we would need something more complex to explain the variety of animals? Darwin's theory of selection is now accepted as such a means, despite it's "apparent simplicity". Similarly the mechanism of operant conditioning can be seen as sufficient to account for very complex forms of behaviour in a wide variety of circumstances without appealing to unproven, non-data driven speculative theoretical approaches like those of Noam Chomsky and others.

Language learning is clearly the most complex task any of us will ever undertake. Yet, the process appears to be considerably less painfull than acquiring an understanding of calculus, or organic chemistry. Noam Chomsky has long argued that this paradox is best explained by the view that humans, and in particular children, have innate abilities that support the acquisition of a language. It is clear that we seem to be naturally good at it, especially before we reach puberty. Moreover, we appear to have a natural need to fill our world with language; in the absence of formal language tutoring a form of language structure develops anyway. Some say a specialised language faculty seems to aid in this process, one that includes innately specified constraints on what forms are possible. These innate, language-specific, information processing mechanisms may be encapsulated in language module of the brain. However, these innate faculties are inferred, hypothesised explanations with no foundation in fact. No biological location has been found for them, no genetic location, no brain structure. It is all inference.

All human languages, even spontaneous ones, show many common principles of language acquisition as well as rules of grammar. The concept of universal grammar proposes that this is due to a set of innate rules, which guide how we acquire language and how we construct valid sentences in it. It thus attempts to explain language in general, and not simply describe the construction of any one specific language *per se.*

There are many alternative theories of human language consistency besides the speculative theory of "universal grammer". One theory is that human environments possess common structures and human language simply responds to the commonality of the world. Universal grammar is a speculative, unproven hypothesis that is still awaiting confirmation and has no evidence to support it other than "rational argument".

Language Acquisiton. At very young age, we acquire our native language by listening to, guessing at the meaning of, and imitating the symbols used by tutors around us. Moreover, during these early years we learn to walk and talk without any explicit need for understanding how we are doing what we are doing. In this process we seem to be helped by a set of Inherent learning strategies, the ability for optimised pattern perception of common, ambient symbols. Infants are exceptionally broad in their abilities to perceive sound qualities. In fact, as infants we can distinguish many more language sounds than we can as adults. During the first year of life, infant brains are actively engaged in optimising acoustic perception for the language sounds that surround them. Such early acquisition of information about native language depends on perceptually mapping both the critical aspects of language, and statistical properties of speech.

It is now clear that infants perceive the various phonetic units, track the frequency of different formants, and extract the boundaries of words from running speech. Patricia Kuhl suggests that language acquisition is based on a combination of factors to provide a powerful discovery procedure for language. Evidence suggests that initial perception parses speech in a universal way in all human infants. Infants have inherent perceptual biases that segment phonetic units without providing innate descriptions of them. They were able to parse and discriminate a wide range of basic phonetic units. Adults, in contrast, are only able to discriminated phonetic units that occur in their first language, but fail to distinguish those that are not used there. Japanese adults, for example, fail to discriminate phonetic boundaries of *r* vs. *l*, boundaries that do not exist in Japanese. Such discrimination is based on general auditory processing mechanisms, rather than on innate phonetic feature detectors for speech.

Language learning requires mapping these probabilistic patterns into language strategies. As infants detect frequency patterns in language input they identify higher-order units. Infants thus discover the critical parameters and phonetic dimensions of the sounds used in their native language. Sensory processing becomes optimised by experience for enhance perception of the specific language around them. Vocal learning unifies language perception and production where vocal learning depends on a comparison of one's own vocalisations to those of others. Imitation forms the integral bong between the perception and production of language abilities and together they become optimised for the first language. If a second language is learned lateron, it will carry the accent typical for the speech motor patterns of their primary language, even following long-term instruction. Similarities in infant-directed speaking styles (increased pitch and exaggerated stress) enhances language learning by assisting infants in discriminating phonetic units, as well as by capturing attention.

Broca's Area underlies the ability to produce speech, but it is not critical for understanding language. Patients with damage will fail to form words properly, and speech is halting and slurred. Wermicke's Area is essential for the ability to understand language. Patients with damage to this area can speak clearly, but the words make no sense (i.e., word salad). The Arcuate Fasciculus connects these two areas. Damage to this connection causes conduction aphasia where language is understood, but neither can words be repeated, nor does own speech make any sense. Capabilities for speech are not distributed evenly across the two halfs of the brain.

Speech is only disrupted when amobarbital is selectively used to anesthetise only the half of the brain which contains these speech centers. Imaging techniques (e.g. fMRI) have identified that bilingual individuals utilise an overlapping set of neurons in the language areas for these two languages. In contrast, individuals who have acquired a second language later in life will likely rely on separate neuronal areas in these speech centers. Late bilungual speakers are also less likely to show strong lateralisation of speech function. This suggests that as two language systems are learned together early-on, they can share the same brain centers without causing catastrophic interference. In adult learning, the best sites of brain real estate have already been taken up by the first language, thus, any new language learning must coopt 'new' territory adjacent to it or on the other half of the brain.

Bird Song

Birds communicate information about danger, food, sex, group movements and many other purposes via acoutic signals. A subset of these have been termed song, as they frequently feature with extended, tonal, melodic characteristics. The Zebra Finch's song, for instance, includes several introductory notes followed by a string of syllables within an extended melodious pattern. Sonograms (i.e., a plot of the intensity of pitch against time) are commonly used as a primary tool for studying and comparing bird songs.

Respiratory muscles force streams of air from large air sacs through the bronchi. Membranes in the syrinx vibrate as air expressed from bronchi passes over them. Syrinx muscles for left and right sound producing structures can act independently, and many birds are able to sing harmonies with themselves. Song appears to play a role in advertising for sex and species recognition as song complexity frequently coincides with the presence of ornate plumage. It also stimulates and synchronises courtship behaviour, stimulates reproductive readiness in females, and contributes to pair bond maintenance. Local song dialects exist in many species.

Successful song in most adult male songbirds depends on memorising the calls of a conspecific tutor during an earlier, sensitive phase in life. The appropriate song repertoire is acquired in a series of distinct stages. Young birds, during an early Sensory Phase, listen to a conspecific tutor and thereby obtain information about the characteristics of its own song. Only a very specific subset of surrounding songs is actually accepted as suitable, suggesting the presence of an in-built song template. Following this sensory phase, young birds actively begin to vocalise themselves. Their Subsong is an atonal, noisy, meaningless repetition of sounds, which lacks recognisable syllables. Akin to human *Babbling* birds practice coordinated movements of the respiratory system, sound producing organs, and related structures (e.g., tongue). During Sensory-motor Phase, young birds spontaneously produce *Plastic Song,* consisting of vocalisations with distinct syllables and recognisable elements. Such "work in progress" will include elements from the song of tutors and elaborate them into a variety of syllables and phrases that even exceed what eventually will be used in its adult song. The ability to hear its own vocalisations are critical for normal development.

In transition to the Mature phase, birds adopt a *Crystallised Song* with syllables and syntax structure that is characteristic of its species. Once established, these song patterns remain fixed in many species, are no more disrupted by deafening, and are presented intact during each subsequent breeding season. In contrast open-ended learners (e.g. starlings and canaries) retain the capacity to adjust or alter their song throughout life.

Song production is under the control of multiple hormonal systems from embryonic gonads. Injections of testosterone induce adult males to sing, even out of season, while similar injections in females have no such effect. The presence of estrogen during male development appears to be essential. When estrogen is blocked experimentally in developing males, testosterone injections fail to elicit song. However, when estrogen had been delivered to developing females, injection of testosterone elicited song in them.

Several neural centers with a role in song have been identified. The *Higher Vocal Center* (HVc) is a group of neurons in the forebrain that is larger in (singing) males than in (non-singing) females. Damage to it blocks song production in adults. The *nucleus of the archistriatum* (RA) in males is larger than in females and its neurons increase in size and dendritic arborisation during song learning. Damage to this area blocks song production in adults. The *lateral magnocellular nucleus of the anterior neostriatum* (LMAN) is neither sexually dimorphic nor shows seasonal change in neuron size or number. Its ablation in young birds interferes with song acquisition but its ablation in adults brings about few deficits as long as song had already been been learned prior to damage. Area X of the paraolfactory lobe (Area X) is sexually dimorphic and new neurons are added in song learning. Damage to it interferes with song acquisition in young birds but not in adults.

As in every behavioural system, a series of independent questions can be addressed for song behaviour in Zebra finches (*Taeniopygia guttata*).

— *Proximate Causation*: Zebra finch song production requires the flow of air through semi-independent vibrators in syrinx and vocal tract. The presence of song, and song repertoire size are reflected in sexual dimorphism of its controlling brain areas and nuclei. A *learning pathway* esists separate of a *motor pathway*. Singing, which is largely restricted to males, is under the control of androgens.

— *Ultimate Causation*: Song in Zebra finches is a learned vocalisation used during courtship and defense of a territory. Advertising the individual's presence it serves to elicit mating opportunities from females and to stimulate the partner's reproductive behaviour and physiology. Moreover, it functions to claim a territory and to repell competitors from it.

— *Phylogeny*: Virtually all 9000 species of birds have the ability to vocalise, including crows, turkeys, owls or nightingales. A large subset of them, including the zebra finch, are characterised by complex vocal organs, distinctive brain circuitry for song, and acquisition of species-characteristic vocalisations through learning. Taxonomically these are all restricted to a single order - the Passeriformes.

— *Ontogeny*: The emergence of adult zebra finch song illustrates the interactions of genetic and environmental factors in behavioural development. After periods of listening to the songs of tutors, starting its own partial vocalisations, rehearsing and adaptating its own song, the species-specific adult version slowly emerges. Song circuits exhibit extensive plasticity even in adults with ongoing neurogenesis and seasonal changes in neuronal morphology.

White-crowned sparrows (*Zonotrichia leucophrys nuttalli*) males sing a single song that shows considerable geographic variation in the form of stable dialects. Bilingual and blended strategies exist at the boundaries. The distinctiveness of the song depends on patterns of natal dispersal and the timing of learning. Subject to reinforcement by the song of neighbors, the system is highly dependent on auditory feedback. The work by Peter Marler, Doug Nelson and others for over 30 years illustrates how genetic and environmental factors interact during the development of a complex communication system.

Cross-fostering experiments illustrate the role of auditory templates in song learning. Young birds reared in the presence of taped song will learn and present that song, even if the tape came from another species. A Song sparrow raised with a swamp sparrow tape will experience little difficulty to learn the swamp sparrow song. Birds in Isolation experiments are raised without access to intact adult song (i.e., no template) and will subsequently show deficiencies in their own song upon maturation. The song does nonetheless contain valid elements of intact adult song. Moreover, Deafening experiments, which deafen birds at hatching, results in song that still contains some valid elements but is an even cruder version than those of isolated birds. When Song preference experiments present young birds with a wide range of conspecific and heterospecific songs, they recognise and preferentially learn the song of its own species. Birds raised in Mixed syllables experiments in the presence of a mixture of swamp- and song sparrow syllables, will acurately produce these syllables in their song but lack the normal adult syntax.

Castration Experiments have shed light on the roles of hormones in song learning. Swamp sparrows that are castrated early in development have low testosterone levels compared to their male siblings. They acquire song but progress to plastic phase only. Treating such birds with injections of testosterone (Enhanced Testosterone Experiment) immediately crystallises the song. Interfering with testosterone function in adult birds (Reduced Testosterone Experiment) degrades previously crystallised song back to plastic.

Brown-headed cowbirds (*Molothrus ater*) are gregarious birds that follows cattle herds. Brood parasites that are raised by parents of different species, no consistent, conspecific tutor available. So, how do they learn their own conspecific song? One of the strongest stimuli is the bird's own crystallised song and feedback from females is important (i.e., action-based learning).

Bird Song vs. Speech Learning

To learn their language, humans and white crowned sparrows follow similar steps, recognition, practice, and clarity. The best time for a human to learn their language is from toddler to twelve years old (Macdonald) and the best time for a white crowned sparrow to learn its song is between 10-50 days from when they were born. Both genes and environment play roles in determining the way each communicate.

The first step for humans and white crowned species in learning their language can be considered recognition. White crowned sparrows and humans listen to a tutor before they begin to communicate their language. The tutor teaches the language and the particular dialect, according to the area. Just like humans, the sparrows also have a dialect depending on where they live. A white crowned sparrow has certain genes that only allow them to learn its own species' songs.

Experiments have been performed in which the sparrow was kept in isolation and played tapes of other song sparrows, the white crowned sparrow will not imitate the other species' song, but will sing an odd song unlike either species' song. Humans also cannot communicate with each other using a different species' language. A study was done in Avignon, France that observed children who were brought up by wolves. It was observed that the children spoke no language at all.

Before studies of this kind were done it was thought that the children might learn to communicate with the wolves, just like the Tarzan story. This is not true, the children could not speak the human language nor the language of the wolves. It is a combination of environment and genes that tell white crowned sparrows and humans to learn their proper language. Genes tell the species to only learn the language of their own species and environment plays a factor in determining which dialect each will use.

Once the white crowned sparrow or infant has recognised its species, song, they can begin to practice the language. Infants first communicate by making sounds. The sounds can range from grunts to sounds that mimic their surrounding environments. For example, before babies can say words, they might say moo- moo when looking at a cow. After sounds come words. By the time the infant reaches one year, they should be making sentences out of words (Macdonald). The white crowned sparrow also does not immediately master its song. The sparrow will first sing a short subsong derived from the tutor's full song. The sparrows keep practicing their subsong just as infants practice their words. After the sparrow has mastered its subsong it can start to form a full song. In both species it is necessary for them to hear themselves in order to vocalise their language correctly.

In conclusion, both white crowned sparrows first listen and recognise their particular species' language. A white crowned sparrow has a tutor to teach the song. Infants usually have parents that teach them their verbal language. Then they must practice the language, starting off slow with sounds and building up. Both species will continue to improve and clarify their language throughout their lives.

MEMORY

Memory, in its broadest sense, refers to an individual's ability to retain learned information. The formation of memory traces is a complex task which all organisms appear to be able to do. Fundamentally, such an ability requires a series of distinct steps: Information must be encoded as meaningful associations are assessed, the resulting information must be recorded in some form, and the memory record must be retrieved when needed.

As humans exhibit little conscious awareness of mental processes, such as memory, these were long considered inaccessible to objective, scientific study. Inferences about the course of encoding, consolidation and storage of memory thus mostly rely on indirect methods such as testing for retrieval abilities. Consolidation of memory appears to involve independent paths at different time frames. Such evidence derives from the existence of different critical time frames for its formation, and its susceptiblility to disruption with different pharmacological tools.

Sensory Memory

Subjects are able to report great detail about a complex stimulus immedeately following its presentation. This ability forms within a few tens of milliseconds and decays again rapidly within a few hundred milliseconds.

Short-term Memory

Short-term memory allows the recall of something from several seconds to as long as a minute. Its strength does appear to depend primarily on attention and not rehearsal. It is thus highly vulnerable to disruption when attention shifts elsewhere. The amount of information that can be held is on the order of 4-6 numbers. This amount can be boosted by grouping them into distinct chunks as for telephone numbers.

The ability to recall such information is contingent on transient patterns of neuronal activity in regions of the frontal and the parietal lobe.

Long-term Memory

This type of memory, lasting hours to months, critically depends on a transfer of the information from short term memory using repeated rehersal.

The hippocampus appears to be an essential structure in such routing. Sleep is thought to improve the consolidation of information, possibly by hippocampal replaying of activity from the previous wake period. Electrical activation of hippocampal circuits reportedly are linked to feelings of Deja Vu.

Declarative Memory

Declarative memory refers to the ability to become conscious of, or declare, facts and experiences. It is also referred to as Explicit memory, when it involves direct recalling information that had been obtained from the external world. Representing knowledge of standard textbook material or events, it is best formed by actively recalling the material in spaced intervals. Compared to other forms or memory it is more volative; more easily formed and more easily forgotten.

The primary neural basis appears to reside in the medial temporal lobe. Bilateral damage to this area results in anterograde amnesia, as in the famous case of the anonymus memory-impaired patient HM. A surgical procedure for epilepsy left him with damage to his brain in the medial temporal lobe on both sides. the hippocampal formation, parahippocampal gyrus, the entorhinal cortex, and the amygdala. He subsequently suffered from severe anterograde amnesia, where transfer of new events into long-term memory was impaired. He was unable to recall events once his attention had focused elsewhere.

Semantic Memory is the ability to consciously recall knowledge of facts that are independent of a specific time and place. medial temporal lobe, diencephalon. Episodic Memory refers to the ability to explicitly recall information about a specific event that has occurred at a specific time and place, medial temporal lobe, diencephalon

Nondeclarative Memory

Implicit memory, which does not need to involve conscious awareness in the act of recollection. Procedural Memory regards the learning of motor skills and habits. Formation requires repeated reinforcement, repetition and practice over many trials rather than recollection. Once formed it is less likely to be forgotten. It is also less easily transferred to related tasks than declarative learning.

Striatum, basal ganglia, Deficits can be assessed using a serial reaction time (SRT) task, backwards reading, mirror drawing, porbabilistic classification, artificial grammar learning, or prototype abstraction.

Motor Responses with Classical Conditioning, Cerebellum

Emotional Responses with Fear conditioning involves an organism's ability to acquire fear responses to a previously neutral stimulus. This occurs when it becomes paired with an aversive stimulus, such as a shock or loud noise. James Wattson's Little Albert experiment illustrated that children learned fear of objects when encounters with them where paired with loud noises.

Priming refers to a context-dependent activations of clusters of neocortical neurons. As they become more activated, they are more likely to come into consciousness.Even reflex pathways are capable of surprising plasticity. For example, circuits that control motor patterns for walking will be subjected to optimisation with sensory input.

Long-lasting Memory

Memory mechainsms critically depend on changes in synaptic functioning. Hebb Synapses and their role in classical conditioning. If a synapse succeeds at driving a postsynaptic neuron above threshold, its subsequent effectiveness is strengthened. Long-term potentiation is based on changes in neural signals which will potentiate a neural response for 1-2 weeks. potentiation involves increased release of the excitatory neurotransmitter glutamate. During habituation, repeated stimulation of a sensory neuron leads to a smaller activation of the postsynaptic motor neuron. The primary cause for this is a progressive reduction of Ca++ inflow into the presynaptic terminal, decreased transmitter release, and a smaller activation of the postsynaptic target. In sensitisation as a result of electrical head shock, release of serotonin activates second messenger systems and phosphorylation of key target molecules. Long-Term Memory depends on new protein synthesis and the formation of new synaptic connections.

HUMAN COGNITIVE DEVELOPMENT

Nature, Nurture, and Cognitive Development from 1 to 16 years: In a Parent-Offspring Adoption Study, Robert Plomin and his fellow researchers conclude that genetics, not environment, play a larger role in the development of the cognitive processes. These researchers are interested in cognitive ability (the mental process of knowing, including aspects such as awareness, perception, reasoning, and judgment). Results for children aged 1-16 show that adoptees, although resembling their adoptive parents minimally at first, become more and more like their biological parents and less like their adoptive parents as they age.

By adolescence, there is a strong resemblance to the cognitive functioning of the biological parents, to the same degree as in the control families, and no similarity to that of the adoptive parents. Despite the fact that the children do not spend any time with their biological parents, they are more similar, cognitively speaking. It suggests that the genes affecting cognitive ability are not all expressed until adolescence and, further, that whatever environmental factors may affect cognitive functioning are not at all correlated with the cognitive ability of the parents who raise the child. The research team conducted their 20-year longitudinal Colorado Adoption Project (CAP) in Denver, CO beginning in 1975.

The CAP followed 245 adopted children and their biological and adoptive parents, as well as 245 matched nonadoptive children and their parents. They recruited pregnant mothers prior to labor. A fifth of the biological fathers were tested, whereas previous studies had not tested the biological fathers. The children were placed in homes at twenty-nine days, and in adopted homes at seven months. The researchers also had a control group of families. The children were matched to adopted families relatively similar to their own based on gender, the number of children in the family, age of the father, occupational status of the father, and father's years of education. Also, the variability in socioeconomic status was representative of the U.S. population. The biological mothers were usually tested during their third trimester and the adoptive parents and control parents were usually tested during the child's first year. Collectively, the thirteen tests lasted three hours long and tested the cognitive abilities of each individual. The tests were given periodically to the children throughout their development and maturation.

CAP also presents a disclaimer stating that they do not believe that their examinations are fully accurate due to exceptions that they did not test, such as mental retardation. They also only represent the middle ninety percent of the U.S. population at the time in which their study began. This study shows that genetic effects on cognitive abilities are not manifested only in adulthood, but also to a considerable extent in adolescence to a lesser degree in childhood, and only slightly during infancy. The different degrees of manifestation attest to the fact that not all genes regarding cognitive ability come arise until the time of adolescence. It also shows that genetics, not environment, contribute to a individuals abilities.

REFERENCES

Cranston, A. 1994. "Self-Directed Animal Development Digestion." *The Journal of Animal Education* (Vol. 65, No. 6). Ohio State Press.

Merrin, L. 2007. *Animals: A Comprehensive Guide* (3rd ed.). Buffalo, NY: Bose-Jarman.

Menzer, J. 1991. *Animal Learning in the 21st Century*. Buffalo, NY: Bose-Jarman.

Roberts, D. 1996. "Facilitating Transformative Behaviors in Animals." *Animal Education Quarterly* (Vol. 47, No. 1). PETS Social Science Publishing.

7

Proximate Mechanisms in Animal Behaviour

An organism may be selected by Darwinian processes to tend towards a particular behaviour, and in the context of this behaviour's evolution, it makes sense to implement this behaviour under certain conditions. But then the conditions change. When the conditions change, evolution may eventually cause a change in how this behaviour works, but in the meantime the behaviour becomes (at least some times) "inappropriate."

Since there is a difference between the ultimate (evolutionary) fitness-related reasons for the behaviours existence and the specific neurological or hormonal mechanism for the behaviour's implementation, the former can make sense in the large view, and the latter often not make sense in the immediate instance.

In the so-called "environment of evolutionary adaptiveness" men are usually with other men who are their close relatives. In humans, it is more often than not the female who moves out of her natal (birth) group to join her new mate in his natal group. Thus, a given man is related to the other men in his group. The other men may include his father, uncles, brothers, and sons. Children in the group are all his children, his siblings children, or otherwise related. Thus, at any one moment in time, if you can give up your life to save every single person in you group, you might be doing something that makes sens in an evolutionary framework. At the very least, there should evolve a mechanism that allows for the option of suicidal altruism.

The guy in the army is not hanging around with his relatives. He is unrelated to his fellow soldiers. But in the evolution of this trait, there did not develop a mechanism to assess this degree of relatedness to group members. The degree of relatedness was an already extant, expected, reliable thing that made up part of the context in which suicidal altruism might evolve. Perhaps all it takes is a sense of "brotherhood" to cause a male's brain to shift into "might-have-to-die-to-save-genes" mode. Certanily, training, living, and fighting together could facilitate this sense of brotherhood among men in the military, and it is clear (and this has been studied) that there are training techniques and other aspects of military life that enhance this phenomenon.

Thus, evolved traits including behaviours always have to be understood, explained at multiple levels. One way to do this is to separately consider the ultimate (evolved, fitness-related) and the proximate (mechanistic) levels as separate. There are other levels but that is for another time. The proximate mechanisms can be very powerful, e.g., Hunger and sex drive.

This chapter explores the most immediate biological functions that are responsible for producing a particular observation. The material aims to provide an understanding of how a behaviour is elicited and coordinated. It includes a search for the impact of genetic components, physiological mechanisms, environmental conditions needed, and hormones on behaviour.

NEUROSCIENCE

Neuroscience, the scientific study of central and peripheral nervous systems in biological organisms, aims to explain behaviour in terms of the activities of the brain. It explores how the activity of millions of individual nerve cells produce behaviour, consciousness and the mental processes by which we perceive, act, learn, and remember. Individual subdisciplines may focus on the structure, function, evolutionary history, development, genetics, biochemistry, physiology, pharmacology, and pathology of the brain.

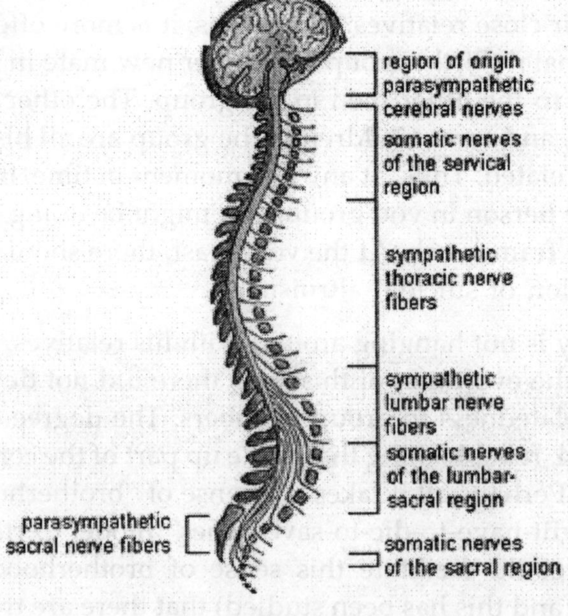

Figure 1. Human nervous system

Neurons

The basic functional unit of the nervous system are neurons, i.e., cells that are specialised for processing and communicating information. Neurons have a basic structure of:

- The cell body (or soma- from the greek word s?μa, plural: somata) is the bulbous end of a neuron, containing the cell nucleus and much of the cells metabolic machinery.
- The axon carries information away from the cell body and may range in size from several microns to as long as several meters in giraffes and whales. Axons may branch into terminal buttons at its end.
- Dendrites receive neurotransmitter secreted by the axon of other neurons. They are sometimes themselves involved in signal transduction.
- Synapse is a specialised junction between cells of the nervous system which permits signaling between them. It consists of the presynaptic and the postsynaptic element that are met in the synaptic cleft. There exist two types of synapses in the mammalian CNS: electrical synapses, through which fast electrical signals are rapidly transmitted between cells and chemical synapses, in which the electrical signal of the presynaptic element is transmitted as a chemical molecule(neurotransmitter)before it becomes again an electrical current inside the postsynaptic cell.

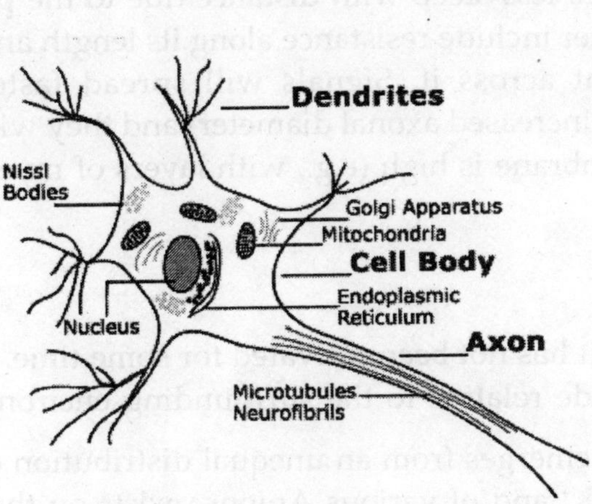

Figure 2. Structure of Neuron

Neurons are able to conduct and process signals that involve small changes in electrical voltage generated across the neuronal membrane. At rest, the inside of the neuron is negative relative to the cell surround—the resting membrane potential. Individual signals are integrated across different regions of the neuron and throughout sets of neuronal networks in order to determine each unit's response and output. A special type of such voltage changes features a rapid spike in voltage resulting from changes in ion conductances—the action potential. Such signals are predominantly transmitted along axonal lines where neurons interact with other cellular entities over greater distances.

Ohm's Law

Basic familiarity with key concepts in electricity is essential in order to appreciate the neuronal mechanisms active during such intergration. Ohm's law holds that the rate at which electric charges flow (i.e., Current or I) depends on the force and its direction exerted onto the charged particle (i.e., Potential, V) and the ease with which the flow can occur (i.e., Conductance, g). The latter can also be expressed as the reciprocal of the conductance, namely as the degree to which the conductor obstructs the flow of charges (i.e., Resistance, R). The flow of electricity is defined as flowing from negative to positive.

$$\text{Ohm's Law:} I = g * V$$

The current is carried by ions moving across the membrane at ion conductances when specific channels allow them to pass. Ignoring its active properties, the axon can in electrical terms be viewed as an insulated cable. An electric potential is able to spread passively along any stretch of membrane. In the process its strength decays and the slope of on- and offsets becomes less steep with distance due to the passive cable properties of the membrane. The latter include resistance along its length and both a resistance and a capacitance component across it. Signals will spread fastest when longitudinal resistance is low (e.g., via increased axonal diameter) and they will spread furthest when resistance across the membrane is high (e.g., with layers of myelin for added electrical insulation).

Resting Potential

A neuron is at rest when it has not been activated for some time. At that state it exhibits a negatively charged inside relative to the surrounding encironment.

The Resting Potential emerges from an unequal distribution of ions across the cell's membrane. An excess of K^+ and of various Anions- exists on the inside while Na^+ and Cl^- exist in higher concentrations on the outside. The membrane is dotted with proteins that permit the flow of ions (i.e., ion pores) or control their flow (i.e., ion channels) across it. In a cell at rest only potassium ions (K^+) are able to flow freely between the two compartments. Initially we observe a net flow of positively charged potassium ions from the inside where they are more numerous to the outside where their concentration is much lower (i.e., concentration gradient). With a net flow of positively charged ions from the inside across the membrane, the inside becomes increasingly negative relative to the outside (i.e., charge gradient). The resting potential settles into an equilibrium when as many potassium ions are pushed out of the cell along their concentration gradient, as will enter the cell along the accompanying electrical gradient. This equilibrium potential for potassium can be calculated using the Nernst potential for known inside and outside concentrations of any given ion

$$\text{Nernst Potential (E)} = 2.303 * {}^{RT}/zF * \log([\text{ion}]^{out}/[\text{ion}]in)$$

where R = the gas constant (8.3143 joules/mole-degree); T = absolute temperature in degrees Kelvin (310 degrees); z = ionic valence; F = Faraday's constant (96,487 coulombs/ mole).

Depending on their respective inside and outside concentrations at the particular cell, different ions will produce different equilibrium potentials (reversal potentials). The ion concentrations are usually similar to those listed:

Ion	*[ion]in*	*[ion]out*	*E*
K^+	400	20	-75mV
Na^+	50	440	+55mV
Ca^{2+}	0.0001	125	+155mV
Cl^-	9	100	-65mV

The Nernst potential for potassium thus can be simplified to the following formula.

$$E_k = 58mV \times \log \frac{20}{400} = -75mV$$

and for sodium is:

$$E_{Na} = 58mV \times \log_{13} \frac{440}{50} = +55mV$$

The cell's resting membrane potential combines all relevant ion currents with K^+ ions figuring most prominently due to their high resting conductance. In addition a small number of Na^+ ions leak into the cell. The resulting resting membrane potential is thus slightly lower compared to the EK^+ at around -65mV. To maintain the concentration gradient across the membrane of resting neurons, ATP continually supplies energy to the Na^+/K^+ ATPase (ion pump) as ions leak across the axomembrane.

Action Potential

An Action potential represents a transient inward current of cations (Na^+). It is generated when the graded membrane potential rises above threshold either spontaneously or because of depolarising input.

During the action potential, voltage-gated sodium channels are opened when the potential rises (i.e., depolarises) from resting potential (-65mV) to the cells threshold (-55mV). These channels close again when the potential further rises to +10mV. As sodium ions spread to neighboring areas of axonal membrane, they open sodium channels there via depolarisation and the signal runs the entire length of the cell (i.e., the all-or-none-principle).

A Refractory period follows the action potential during which time the the neuron reestablishes its normal resting potential. At the begining of this period it is impossible for another signal to be transmitted, this is called absolute refractory phase. This is followed by the relative refractory phase where it is possible to send another signal but more exitation than normal is needed.

The Duration of an action potential is 1-2ms in Vertebrates and 1-100ms in Invertebrates. Frequency of firing ranges from <1 to about 100/sec (100Hz). The Amplitude ranges between 70-80mV when recorded intracellularly and 5-200μV when recorded extracellularly.

Synapse

Electrical Synapses

Two electrically excitable cells, such as neurons or muscle cells, may be electrically coupled where an action potential in one cell moves directly into the other via arrays of gap junctions. Electrical synapses are fast but cannot be modulated. They are mostly used in neuronal circuits for escape behaviours where speed of conduction is essential. Electrical synapses (gap junctions, electrotonic junctions) allow current to flow between separate neurons when ions pass through gap junctions. Connexons (where 6 connexin proteins form a hemi-channel) are the actual pores that allow ions to flow past the two membranes. A connexon in the presynaptic membrane lines up precisely with its respective equivalent in the postsynaptic membrane, forming a continuous channel from one neuron to another. With a pore diameter of about $1.5m^{-9}$ many small molecules can pass through efficiently.

Intracellular Ca^{2+} concentration, pH, or phosphory-lation of connexins can profoundly alter the easy with which ions annd proteins may pass through the pore. As there is no synaptic delay in transmission of current from cell to another, the conduction of potential changes is considerably faster than through chemical synapses. Although electrical synapses are often bi-directional, some synapses pass current better in one direction than the other (i.e., rectifying synapse) Electrical synapses are commonly used in time-critical processes (escape behaviours), when rapid synchronisation of many cells is needed (e.g., vertebrate cardiac muscle), between glial cells, or early in development.

Chemical Synapses

Cells are also able to communicate chemically across a gap (i.e., synapse), which forms a directional connection from one neuron to another cell. The neurotransmitter is emitted from the terminal endings of one cell;s axon onto a dendrite or body of a second cell.

The arrival of a signal leads to the release of a neurotransmitter from the presynaptic terminal, diffusion across the synaptic cleft, and binding to receptors in the postsynaptic membrane. Ionotropic or metabotropic receptors are binding proteins which alter ion conductances at the postsynaptic cell membrane (e.g., increase in Na^+ conductance depolarises and is excitatory, CL^- hyperpolarises and is thus inhibitory).

When excitatory, the postsynaptic cell is depolarised with an excitatory post-synaptic potential (EPSP) and an action potential is elicited if the threshold is reached; in inhibitory connections, the postsynaptic cell is hyperpolarised with an inhibitory post-synaptic potential (IPSP) and it will thus be harder for other inputs to drive the cell towards an action potential. A single input is rarely sufficient to lead to an action potential in the post-synaptic cell. Multiple EPSPs may add and reach the threshold when a series of action potentials arrive at high rate. Chemical synapses are capable of integrating a complex scenario of inputs.

Neurotransmitter refers to a compound that is released at a synapse and diffuses across the synaptic cleft to act on a receptor located on the membrane of a postsynaptic cell, which may be another neurone, a muscle cell or a specialised gland cell. A wide range of chemcials are used as neurotransmitters in the nervous system. Stored in synaptic vesicles. they are released during the arrival of an action potential and produce changes in the excitability of the postsynaptic membrane. Ca^{2+} influx at the axon terminus is required for synaptic release.

Neuromodulator refers to a compound that is released within a localised region of CNS, the receptor for which is not necessarily sited on an anatomically apposed postsynaptic cell. Thus a neuromodulator may affect several postsynaptic cells with specificity conferred mainly by the distribution of receptors. Main action is on second messenger systems, eg. cAMP or inositole triphosphate, presumably affecting protein phosphorylation

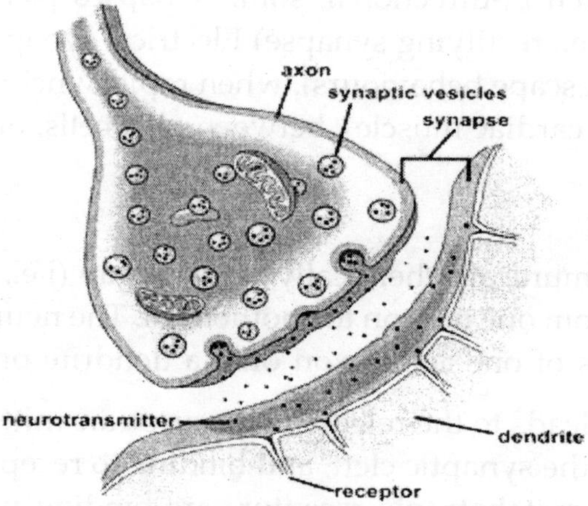

Figure 3. Chemical Synapse

Sometimes the same neurochemical may have rapid transmitter type effects, followed by longer modulatory influences. This suggests that neurotransmitter and neuromodulator effects may be most effectively classified at the receptor level. Activation of receptors on a protein structure directly incorporating an ion channel (a ionophore) are defined as neurotransmission while activation of receptors coupled indirectly to ion channels (eg. via second messanger systems) are defined as neuromodulation.

Signal Strength

For the signal to be passed from one neuron to the next it must have enough energy to break a point called the threshold. Once the threshold is broken the signal is transmitted. The neuron fires at the same strength every time. The strength of a signal is decided by how many different neurons are being fired and at what frequency they are being fired. The ratio of glial cells to neurons in the nervous system is disputed.

Glial cells function as support for the neurons; they produce the myelin sheath which surrounds some neurons and also form part of the blood-brain barrier. The blood-brain barrier is a structure that prevents certain substances in the bloodstream from reaching the brain. Many axons are sheathed with tubes of myelin, which is a fatty material. Myelin is produced by the glial cells. The myelin sheaths on axons have gaps, which are called the nodes of Ranvier. Myelinated sheaths help transmit information quickly and efficiently.

Synaptic Integration

Activity in any presynaptic neuron alone is rarely sufficient to elicit an action potential in the postsynaptic neuron. Summation of synaptic inputs, however, can occur when a neuron receives multiple exitatory inputs in short order. Synaptic processing describes the ways in which the electrical events in different neurons may interact through interactions at their synapses.

Synaptic inputs are mostly located on the dendritic tree of a neuron, while axonal action potentials represent the neuron's output. In order to drive a neuron the integrated influence of the electrical events generated by the synapses must be sufficient to spread through the dendrites to the soma and into the initial segment of the axon (i.e., the axon hillock, trigger zone). The latter is the site of action potential generation. Synaptic integration is thus key to understanding the contribution of individual neurons to information processing within neural networks.

In spatial summation the cell responds when activity from multiple inputs arrives at the same time. If the combined excitation brings the cell to threshold, an actionpotential is initiated. The arrival of a spike train in a presynaptic neuron may produce multiple EPSPs. If these inputs begin to build on each other, the postsynaptic cell may reach threshold and produce an action potential.

Passive membrane properties are important. Long time constants increase the chances for temporal summation. Large space constants determine the likelihood of spatial integration. The size of the postsynaptic cell counts, as a synaptic current delivered into a small neuron will produce a much greater effect than in a large postsynaptic cell.Location of synapses is important. Control of the cell's activity usually driven by excitatory synapses on dendrites. Inhibitory synapses located on the cell body are able to shut down activity. Synapses at presynaptic release sites are modulatorym controlling the amount of neurotransmitters released.

Mammalian Nervous System

The neurons can all be placed in one of two systems, the central nervous system or the peripherial nervous system.

Central Nervous System

The central nervous system has a fundamental role in the control of behaviour. It contains the brain and the spinal cord which are both incased in bone that protects them from mechanical injury. Both the brain and spinal cord recieve signals from the afferent neurons and send signals to muscles and glands through efferent neurons.

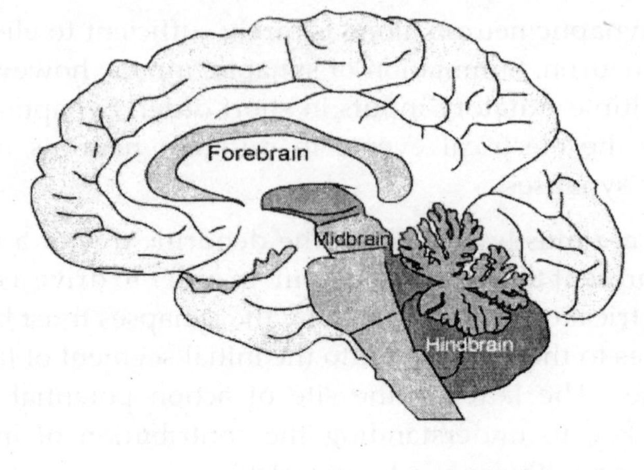

Figure 4. Structure of human brain

Structure and Function

The brain is split up into three major layers, the hindbrain is the first, the second is the midbrain, and the forebrain is last.

Hindbrain

The hindbrain is a well protected central core of the brain and includes the cerebellum, reticular formation, and the brain stem. The cerebellum plays an important role in the integration of sensory perception and motor output. It utilises constant feedback on body position to fine-tune motor movements. The brain stem contains the pons, and the medulla oblongata. The pons relays sensory information between the cerebellum and cerebrum. The medulla oblongata is the lower portion of the brainstem. It controls autonomic functions such as breathing and vomiting, and relays nerve signals between the brain and spinal cord. The reticular formation is a part of the brain which is involved in stereotypical actions, such as walking, sleeping, and lying down.

Midbrain

This part of the brain is located between the forebrain and the hindbrain making up part of the brain stem. All sensory and motor information going to and from the fore brain and the spinal cord must pass through the midbrain.

Forebrain

The anteriormost division of the developing vertebrate brain that contains the most complex neural network in the CNS. The forebrain has two major divisions, the lower diencephalon, which contains the thalamus and the hypothalmus, and the upper telencephalon, which contains the cerebrum.

Brain activity

In the past only two methods of observation were available. The first was observing individuals who have recieved brain damage and assume that the part of the brain that was damaged controlled the behaviour or sense that had changed. The second was connecting electrodes to the outside of someones head and recording the readings.

Newer methods include computed tomography (CT scan), positron emission tomography (PET scan), magnetic resonance imaging (MRI), and superconduction quantum interference devices (SQUID).

Peripheral Nervous System

Any part of the nervous system that is not part of the central nervous system is part of the peripheral nervous system. The nerves in the peripheral nervous system is split up into the autonomic and somatic. The somatic connect the central nervous system to sensory organs (such as the eye and ear) and muscles, while the autonomic connect other organs of the body, blood vessels and glands.

SENSORY BIOLOGY

Behaviour requires that an animal obtains information about its environment. Thus, to understand behaviour we need to understand how organisms perceive their environment (i.e., sensory systems biology, sensory biology). By understanding how your senses gather information, we gain a better and more thorough understanding of our behaviour. Common research questions in sensory biology focus on:

— What is the stimulus (modality of information)? Is it mechanical, chemical, etc.

— How does the sensory system encode the stimulus? Sensory filtering, transformation, transduction, amplification

— Where is this information processed? Neural anatomy, network connections, etc.

Electroreception

Electroreception refers to the biological ability to sense electrical impulses. As water is a better conductor than air, electroreception is more common in aquatic creatures. Individuals utilise this sense to locate living organisms as sources of electrical energy. Species specific differences in sensory reception. Human perception utilises 5 sensory modalities: sight, hearing, touch, taste, and smell. Receptors are classified based on the source of the stimulus. Interoreceptors convey information from within the organism; Proprioreceptors report on the spatial position of body parts relative to one another; Exteroreceptors obtain information about the outside. Subcategories of the latter include Somatoreceptors that capture events on the body's surface and Teloreceptors which monitor stimuli at a distance fro the body.

Sensory selectivity: Refers to the subset of stimuli, which an animal detects and responds to; the Umwelt: Sensory World. Subjective set of stimuli to which an animal is responsive in a given motivational state. Each species has evolved responses only to those stimuli that prove relevant. It is this simpler world that actually falls within the animal's perception at any particular moment.

Sensory Stimuli

Stimulus Energy: There are three main forms of energies that can alter cellular processes and thereby activate sensory systems:

- *Mechanical:* (particle movement—near field, pressure waves—far field, compressional vs. transverse waves): Hearing, Vestibular, Touch (via hair cells in cochlea, statocyst, or semicircular canals) via hair cells in acoustic, equilibrium and lateral-line systems are responsible for the transduction process from cilia that respond to shearing forces and end with processes that depolarise the cell membrane of afferent neurons. Stimulus specificity is provided through Accessory Structures (e.g. tectal mebrane, statolith, and cupula). Touch: routing response in babies, lordosis, grooming.
- *Electromagnetic*: (waves): Light, Heat, Electrical, Magnetic (e.g., Phototransduction, heat sensors). Radiant Heat: thermoreception, Example: Detection of prey in rattlesnakes; Electric Fields: electroreception, Examples: Detection of prey in sharks, communication in electric fish; Light: perception of absolute light levels, color balance, polarisation, Examples: fireflies.
- *Chemical:* (particle movement): Olfaction/Pheromones/Taste (via chemical receptors trigeminal, taste, smell.
- *Odor*: chemoreception, pheromones species-specific odor cues (priming, releasing), Example: silk moths Sound and Vibration: mechanoreception, proprioreceptors, sound window, transmission in water vs. air, boundaries (surface waves), complex content (frequency modulation, amplitude modulation), Example: bird and whale songs, human speech, echolocation in bats and whales, long-distance communication in elephants and whales Movement of particles: Example: spiders.

Pheromones

One of the many ways in which organisms can communicate with each other is through the use of pheromones. An organism generates and emits these hormonal chemicals in order to relay a message to another member of the same species. Ants and bees demonstrate two prominent examples of pheromone usage, which acknowledges their incredible capability to organise the behaviours of the whole colony.

Ants produce numerous different pheromones, each with its own distinct purpose. Ants secrete pheromones to attract mates, to signal danger to the colony, or to give directions about a location. Other pheromones act as deterrents keeping out unwanted ants from foreign colonies or preying insectivores. Still other pheromones communicate ants to congregate. This explains how assiduous ants exercise remarkable cooperation in building a colony.

Pheromones maintain the cohesiveness and organisation of the complex ant communities. Certain types, such as alarm pheromones, produce a "releaser effect", which induces a quick response and may be used to tell other ants to evacuate a dangerous area such as an approaching lawnmower. For example, when a spider approaches an ant will release alarm pheromones that alert all the other ants. Ants may also discharge alarm pheromones as a result from being diverted from their work, e.g. heavy human steps. Releaser pheromones are also used to mark territory. As the chemical deposited dries, it signals to other species members of the territory's occupant.

Other pheromones create a "primer effect" that entices other ants for actions. Such pheromones are useful in mating rituals and only affect ants of the opposite sex. Primer pheromones can send signals to the endocrine system, to make appropriate changes, for instance ovulation required for successful mating.

Biologist E.O. Wilson discovered in the 1960's that the organic chemical for each pheromone varies tremendously depending on what signal it entails. Ants taste and smell a substance that evaporates off the chemical laid down by another ant. Wilson observed in slow motion films that ants do this by moving their antennae from side to side. For example, alarm pheromones are discharged into the air, and expand in a circle of smell. Ants can determine the concentration of the pheromone, and thus determine the proximity of the source of danger.

Bees are also well known for communicating through the use of pheromones. Like ants, bees have a variety of purposes associated with the pheromones, such as marking and behaviour. Honeybees release chemical signals for marking food sources, marking their hive, in scenting potential hive sites, and in assembling swarms for flight. Although each hive has a particular scent, different colonies can be easily integrated during times when honey flow is heavy since the colony's odor is inundated with the scent of nectar.

Virgin queen bees emit a behavioural pheromone released with their feces. When the new queen feels threatened by the workers she uses this pheromone as a repellent.

Another widely identified chemical signal in bees is the Queen Mandibular Pheromone (QMP). This pheromone ensures that the queen is the only reproductive female in the hive by compromising the reproductive systems of worker bees. It also provides an attractant signal to the drones.

Sensory Systems

Sensory systems allow us to form internal representations of our surrounding world, by transducing stimulus energy into trains of neural signals which are conveyed along specific neural pathways. Jelly Bean Example. Hawaiian punch and Cherry have the same color. Visual information about jelly beans is incomplete. By blocking your nose while eating a jelly bean, you prevent smell from providing information. Taste does not allow you to determine which bean is punch and which is cherry. With olfactory information, this decision is easy. Finally, the cinnamon bean activates your trigeminal system. This provides critical information independent of taste and smell.

A stimulus is any form of energy that can be detected by the body. A signal is the physical coding of information (e.g., a message) capable of transmission through environment. Sensory processing includes all central acts of information processing, which link the initial stages of sensory reception with the creation of a subjective sensory percept. Sensation is the neuronal activity resulting from the transduction of stimulus energy into electrical activity (also Sensory processing) includes a series of distinct steps:

- Reception the ability of a cell to respond when matter or a specific form of energy acts upon a <Sensory Receptor>: A cell endowed with the ability to absorb a specific kind of stimulus energy. Stimulus filtering (also stimulus tuning): A receptors responds only to a narrow cocktail of (physical) characteristics.
- Sensory Transduction chain of physiological reactions which convert sensory input into electrical impulses. This process translates the amount of stimulus energy into changes in membrane permeability (e.g., opening sodium channels towards the production of an action potential). Initial receptor responses are often graded and proportional to the strength of the input signal (e.g. membrane permeability of stretch receptors to ions is proportional to the force applied to the receptor).
- Amplification when weak sensory inputs are intensified
- Transmission as input signals are conveyed to the nervous system. The intensity of the graded response determines the frequency of generated action potentials. The rapid depolarisation and hyper-polarisation of an all-or-nothing action potential spike is fairly uniform in amplitude and duration (< 2 msec). Frequency code:

Information transfer based on the rate of action potentials of up to 500AP/s for intense stimuli. Sensory receptors may be neurons who themselves project axons to the CNS or non-neuronal receptors which activate neurons via synaptic (i.e., neurotransmitter) signals

— Integration: Processing of information begins as soon as stimuli are received. Summation of multiple graded responses influences frequency of action potentials.

Sensory Receptors

Sensory receptors are able to respond to a particular stimulus energy (i.e., sensory modality) and transduce it into neural impulses. Tonic receptors respond with a constant rate of firing as long as stimulus is applied (e.g., pain). In contrast, phasic receptors produce a burst of activity during the onset of the stimulus but quickly reduce their firing rate if the stimulus is maintained (e.g., odor, touch, & temperature). With sensory adaptation, the organism ceases to pay attention to a constant stimuli. The adequate stimulus for a praticular receptor is the one that requires the least amount of energy to activate the receptor. Sensory Receptors

Mechanoreceptors

— *Hair Cells*: Hair cells are sensory receptors that transduce physical forces into changes in electrical actrivity. Their name derives from a bundle of cilia on the apical surface of the cell. Embedded bundles of cross-linked actin filaments are anchored to the top of the cell membrane, where they control ion currents across a set of channels. Shearing of the tuft of cilia towards one side or the other, changes the conductance of ions across a set of stretch-activated ion channels (i.e., tip links).

— *Stretch receptors (MRO)*: The muscle receptor organ spans the joint between two adjacent abdominal segments in crayfish. These are sensory organs that provide information about posture and movement of the individual. Each MRO has a thin muscle fibre that runs in parallel to a muscle bundle used by the animal to maintain the position of its abdomen. When the muscles contract or are stretched they cause the MRO to change its firing pattern, providing information to the nervous system regarding the relative position of the abdominal segments. There are two kinds of MROs. Tonic MROs respond to chronic stretch with continuous firing and habituate slowly. Phasic MROs respond to rapid changes in posture, but habituate quickly when the posture is maintained for more than a few moments.

Electro-Magnetic Receptors

— *Light-sensitive neurons:* rods (B/W) and cones (color) in the mammalian retina contain dense stacks of membrane with large numbers of light-sensitive pigments. The latter consist of a protein (opsin) and a bound chromophore (retinal). Retinal is able to capture photons, induce a conformational change in opsin, and thereby activate a G protein coupled second messenger cascade.

– *Electroreceptors:* Electric fish, Sharks, Platypus: Some animal possess specialised sensory cells that enable them to detect changes in the electromagnetic field around them. These organs can be used by predators to locate prey by the electrical activity of their nervous systems and muscles, or in some cases as a means of communication (when coupled with the ability to produce pulses of electricity as seen in some electric fish).

– *Thermoreceptors:* Some snakes hunt their prey using body heat. Heat travels through the the atmosphere as infra-red (long wavelength) electromagnetic radiation, and is detected by cells sensitive to changes in temperature.

Accessory Structures

Interior of a statocyst gravity receptor, an equilibrium organ of invertebrates, such as a crayfish. A fluid-filled vesicle is lined with mechanoreceptors (hair cells). It contains one of more dense objects (eg. sandy or stone-like elements)—the statoliths. As gravity pulls these objects down, activity of sensory cells below them increases when their cilia are sheared. From this, the central nervous system can extract information about the direction of gravitational pull. When the animal molts it loses the lining of the statocyst and with it the sand grains. After molting the animal rebuilds the otoliths with materials from its surrounding.

Sensory Processing

The receptive field of a sensory neuron is the specific region of a sensory surface (e.g., area on the retina) that, when stimulated, causes a change in activity of a neuron

Feature Extraction

Logical gates can be created by driving an action potential in the postsynaptic cell only if e.g., multiple input sources are active concurrently (i.e., spatial summation)

– *Coincidence detection*: Activity in any presynaptic neuron alone is usually not sufficient to produce and action potential in the postsynaptic neuron. Summation of synaptic inputs can occur, however, when a neuron receives multiple exitatory inputs in short order. Spatial summation allows a cell to fire if two inputs are active at the very same time. The window of opportunity during which concurrent events must occur depends on the time constant of the neuronal membrane. For coincidence detectors the time constant must be short; if the cell works as a temporal integrator, it must be long.

When localising a stimulus source, systems for the discrimination of left vs. right are often based on two sub-systems. These are often mirror-images of each other and located to the left and the right of the midline (i.e., Omega neurons in crickets). They are tightly coupled through lateral inhibition, where activation of one side automatically shuts off its contralateral (i.e., opposite side) opponent. Such a design is uniquely able to allow resolution of extremely small time differences in when a sound signal arrives at the ear facing the source than in the one facing away.

— *Contrast enhancement:* Lateral Inhibition: Multiple units with similar characteristics are wired to inhibit each other's activity. The unit that fires first/strongest will prevent all others from firing. (e.g., Discrimination of left vs. right in auditory signals using a pair of Omega neurons in crickets).

— *Measure time-delay between two inputs*: Action potentials (AP) travel along axons at a defined speed. They thus take longer to arrive at the target the further the AP needs to travel. Multiple neurons with similar characteristics are layed out in a longitudinal array. They receive input from two sources one fed in from one side of the array, the other from the other side. Acting as coincidence detectors they respond best when the signals on both inputs match. Each member of the array is most sensitive to a particular time difference.

Example: Vision

The resolution of the retina varies between different regions. A sensitive, b/w, high resolution central region (i.e., fovea) is surrounded by a peripheral field with poor resolution and color reception. No photorecptors are present at the location (i.e., blind spot)) where ganglion cell axons leave the eyeball to project to the brain via the optic nerve.

Systems for the processing of visual information extract features from the visual field at multiple hierarchical levels. The photoreceptor cells in the retina detect points of light (image). The retinal ganglion cells respond to point contrast (form and color). The lateral geniculate nucleus provides a first rapid analysis of high-contrast, black and white features. The primary visual cortex provides a slower but more in-depth analysis with an emphasis on linear contrast, color, depth, object vs. background motion.

Retinal receptor cells (i.e., cones for color and rods for black and white) feed into bipolar on or off cells in the retina Retinal ganglion neurons have center-surround characteristics and thus respond to spots of light of a particular size and in specific places of the visual field. Receptive fields are circular and monocular. Many axons cross to the opposite side as they project towards the lateral geniculate nucleus (LGN) and the primary visual cortex (V1).

The visual cortex is arranged in multiple horizontal layers with outer layers focused on the processing of simpler features. Most neurons respond best to oriented bar stimulus, sensitive to motion, monocular or binocular. As the information proceeds further into the structure, features at increasing levels of abstraction are analysed. Vertical columns across these layers primarily respond to input from one specific eye and respond to a particular feature orientation.

Simple cells respond best to bars of given orientation at given location within receptive field. These oriented edge detection neurons feed their output into motion-sensitive neurons. Complex cells are less sensitive to stimulus position within the receptive field and more sensitive to stimulus motion.

Hypercomplex cells respond like complex cells, but feature in addition an inhibitory region at one end. Some neurons at the highest-level receptive fields are quite specific i.e. a neuron that only responds to faces, faces with particular expressions, or belonging to one particular individual (i.e., grandmother neuron). Flicker-fusion frequency, the lateral geniculate body and dyslexia

The visual parts of the brain strive for hypotheses about the contents of the outside world working in essence with an image of rather poor quality and thereby is able to produce our rich everyday visual experience.

Sensory Perception

Sensory perception is the interpretation of sensory signals within the CN where it produces an internal representation of electrical activity from sensory organs. Specificity of sensory impulses derives from transmission via labelled lines as it is largely dependent on which part of the brain receives the signals.

Pheromones in Ants and Bees

One of the many ways in which organisms can communicate with each other is through the use of pheromones. An organism generates and emits these hormonal chemicals in order to relay a message to another member of the same species. Ants and bees demonstrate two prominent examples of pheromone usage, which acknowledges their incredible capability to organise the behaviours of the whole colony.

Ants produce numerous different pheromones, each with its own distinct purpose. Ants secrete pheromones to attract mates, to signal danger to the colony, or to give directions about a location. Other pheromones act as deterrents keeping out unwanted ants from foreign colonies or preying insectivores. Still other pheromones communicate ants to congregate. This explains how assiduous ants exercise remarkable cooperation in building a colony. Pheromones maintain the cohesiveness and organisation of the complex ant communities.

Certain types, such as alarm pheromones, produce a "releaser effect", which induces a quick response and may be used to tell other ants to evacuate a dangerous area such as an approaching lawnmower. For example, when a spider approaches an ant will release alarm pheromones that alert all the other ants. Ants may also discharge alarm pheromones as a result from being diverted from their work, e.g. heavy human steps. Releaser pheromones are also used to mark territory. As the chemical deposited dries, it signals to other species members of the territory's occupant.

Other pheromones create a "primer effect" that entices other ants for actions. Such pheromones are useful in mating rituals and only affect ants of the opposite sex. Primer pheromones can send signals to the endocrine system, to make appropriate changes, for instance ovulation required for successful mating.

Biologist E.O. Wilson discovered in the 1960's that the organic chemical for each pheromone varies tremendously depending on what signal it entails. Ants taste and smell a substance that evaporates off the chemical laid down by another ant. Wilson observed in slow motion films that ants do this by moving their antennae from side to side. For example, alarm pheromones are discharged into the air, and expand in a circle of smell. Ants can determine the concentration of the pheromone, and thus determine the proximity of the source of danger.

Bees are also well known for communicating through the use of pheromones. Like ants, bees have a variety of purposes associated with the pheromones, such as marking and behaviour. Honeybees release chemical signals for marking food sources, marking their hive, in scenting potential hive sites, and in assembling swarms for flight. Although each hive has a particular scent, different colonies can be easily integrated during times when honey flow is heavy since the colony's odor is inundated with the scent of nectar. Virgin queen bees emit a behavioural pheromone released with their feces. When the new queen feels threatened by the workers she uses this pheromone as a repellent.

Another widely identified chemical signal in bees is the Queen Mandibular Pheromone (QMP). This pheromone ensures that the queen is the only reproductive female in the hive by compromising the reproductive systems of worker bees. It also provides an attractant signal to the drones.

NEURAL CONTROL OF MOTOR SYSTEMS

The neural control of behaviour will be most obvious in cases where the brain performs only a small amount of processing. Alternatively we can explore neural systems where functions are limited to a primary goal and which are highly optimised to achieve it. The latter is particularly true for sensory systems which encode a single stimulus parameter, or motor systems in escape behaviours. We can ask how a particular behaviour is produced, with characterising muscle activities in the behaviour, exploring the neural motor centers that control them, working our way upstream in the flow of information processing, until we reach the mechanisms responsible for sensory input.

Executive Control

Complex and coordinated motor patterns can often be centrally generated without requiring sensory input. Such fictive motor patterns may often even be elicited in reduced preparations and studied in vitro. Such fictive motor patterns arise from bursts of activity in isolation, or as a consequence of circuit interactions.

Command Neurons

Command Neurons are cells, when activated, are capable of producing complex behaviour patterns in the absence of any meaningful external stimuli. Activation of these neurons is both necessary and sufficient. Examples include the production centers for mating song in crickets, the giant fibers in crayfish escape circuitry, or the Mauthner neurons in goldfish. The concept in which single spikes are viewed as sufficient for behaviour may be somewhat dated. There is little evidence that any single neuron is able to exert control over a complex behaviour and its motor programs. Neurons cannot be viewed as single identities, they are always part of larger circuits, they control as well as receive input, and their activity thus only makes sense within the context into which they are embedded.

Central Pattern Generator

Central Pattern Generators (CPGs) are circuits that organise repetitive motor patterns, such as those underlying feeding, locomotion or breathing. The production of rhythmic motor patterns and its control by higher-order command and modulatory interneurons can be explained by the intrinsic membrane properties and connections of these neurons.

Studies concern the specific timing of activation of the component neurons, how sensory neurons alter or gate CPG output, or how motor patterns are activated, inhibited, or modified by modulatory drive. Reciprocal Inhibition of functional antagonists can produce alternating discharges in neurons and its associated motor output. Complex, multidimensional oscillations can emerge from multiple, overlapping mechanisms.

Stomatogastric System of Crustaceans

The crustacean stomatogastric nervous system (STNS) generates rhythmic motor patterns in the stomach and other regions of the foregut. The patterns, controlling more than 40 pairs of striated muscles, serve to grind and filter food using multiple and variable cycles. Over the past 40 years the small section of the nervous system that controls it, has been used to gain an understanding how neural circuit dynamics arise from the properties of its individual neurons and their connections.

The STNS consists of a group of four linked ganglia, the paired commissural ganglia (CoGs), the unpaired esophageal ganglion (OG), and the STG . The STG consists of ~30 motor neurons that move the muscles of the gastric mill and pyloric regions of the stomach. The CoGs and OG control STG activity via descending modulatory input. Accessible to simultaneous recordings of all relevant circuit neurons, the electrical, chemical and molecular properties of individual nerve cells and the direct and modulatory connections between them, can be studied along with the various emerging network properties.

Crayfish Escape

A tailflip consists of a rapid, ventral flexion of the abdomen. It is produced in response to stimulation from a potential predator and results in a backward or upward propulsion that causes the animal to escape from it. The control system for this behaviour is formed by 2 pairs of giant neurons—the lateral giant neurons (LGN) and the medial giant neurons (MGN). The concept of command neurons was used to explain the causation of that behaviour. The actual components of the system are:

- *Bipolar sensory neurons:* tuned to touch or high frequency water movements (>80Hz)—sensory filtering. Sensory neurons produce chemical synapses onto non-giant interneurons, and electrical synapses to giant interneurons subthreshold stimulation triggers non-giant escape swimming, produced by a pattern generator consisting of interneuron networks
- *Medial Giant Interneurons:* brain to last abdominal ganglion, contraction of fast flexor muscles in all abdominal segments, direction of escape: straight back
- *Lateral Giant Interneurons:* segmentally repeated; input in abdominal segments, flexor contraction in anterior segments of abdomen, direction of escape: upwards giant motor neurons and fast flexor motor neurons for flexion of abdomen via giant fibers. Habituation only occurs if stimulation of sensory neurons is sufficiently intense to drive an action potential in the giant fiber system
- Re-Extension of abdomen follows lateral giant tailflips within 100ms of stimulus, reflex of hair receptors and muscle receptor organs
- Excitation of extensor systems via excitatory interneurons with onset of tailflip to prepare for re-extension
- Inhibition of extensor systems via inhibitory interneurons while tailflip is happening then Inhibition of further tailflips while re-extension is happening: inhibition of sensory neurons driving giant fibers, inhibition of flexor muscles

BEHAVIOUR ENDOCRINOLOGY

The body has two types of glandular systems, the endocrine, which generally secrete hormones through the bloodstream, and the exocrene which secrete fluids to the outer surfaces of the body, such as sweating.

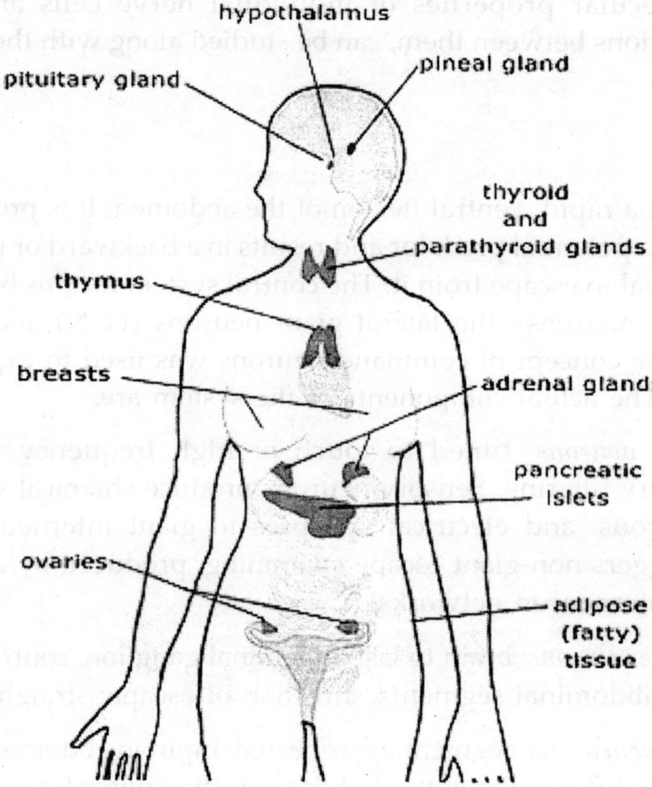

Figure 5. Human endocrine system

Endocrine Glands

The Endocrine System combines neural and glandular mechanisms which control physiological functions/behaviour via the secretion of hormones. Hormones are chemical signaling molecules which play an integral role during development (organisational effects) and day-to-day functioning (activational effects) of target tissues at critical times. Secretory cells of a particular type are often clumped together into a well defined gland (e.g. pituitary, thyroid, adrenal, testes, ovaries). Secreted at that site they distribute throughout the body via the blood stream, and cause physiological changes at any other sites.

— Steroids: derived from cholesterol

— Amines: derived from amino acids

— Peptides: short chains of amino acids via protein synthesis

Sex hormones, largely steroids, are secreted from gonads and adrenal cortex. Androgens (e.g., testosterone) are usually higher in male mammals while levels of estrogens (e.g., estradiol) in female mammals exceed those in males. Circulating levels of sex hormones then provide the basic organisation for gender phenotypes.

Neurohormone refers to a compound that is released into the bloodstream at specialised neurohemal release sites. It binds to receptors anywhere in the body and thereby coordinates disparate biochemical responses.They are released from glands, transported via the circulatory system and influence the activity of target organs.

Functionally hormones are categorised as Effector hormones (e.g. Vasopressin, Oxytocin) or Tropic hormones, releasing factors (e.g. Gonadotropin Releasing Hormone—GnRH, Growth Hormone Releasing Hormone—GHRH). Target Organs receive hormones via blood stream, respond directly or release their own hormones in response (steroid hormones), and these hormones circulate back to turn off hormonal secretion: endocrine feed back loops. Compensatory hypertrophy results from feedback loops control levels of activity where systems are upregulated until they achieve sufficient functional effects (e.g. thyroxine and goiter).

Sex Determination

Hormonal influences during critical periods produce fairly permanent changes in nervous and endocrine systems. Sex-determination may be chromosomal or environmental:

— Mammals: hormonal cascade of events takes an individual down one of two paths (default is female). testes determining factor (tdf) in mammals or Sexchromosomal Abnormalities: Turner syndrome (XO monosomy), Klinefelter syndrome (XXY trisomy), XYY syndrome ("supermale"), Multi-X syndrome ("superfemale"), XX Male syndrome (SRY gene transference) XY Female syndrome (SRY gene missing)

— Birds: Males are homogametic (ZZ), females are heterogametic (ZW)

— Turtles: temperature

— Fish: social stimuli

Internal sex organs: Precursors for both internal sex organs are present in the embryo: the Mullerian system for the female sex; the Wolffian system for males. Controlled by the levels of circulating hormones, only one set of organs develops while the other shrinks. Secreted by the testes, Mullerian inhibiting hormone (i.e., a peptide hormone) has defeminising effects while androgens (i.e., steroid hormones) exert a masculinising effect. Gonads: testes or ovaries, first to develop. External genitalia: Androgens are essential for the development of primary sexual characteristics.

Androgen insensitivity syndrome: A genetic male lacking functional androgen receptors develops testes, secretes Mullerian inhibiting substance and androgens. Internal sex organs progress onto the intended male path, but female external genitalia result from the inability to respond to androgens correctly. Turner's syndrome, with genotype XO (i.e., a single X chromosome and no Y), lacks testes and ovaries.

Maternal effects explain influenced by effects of the mother that are not due to direct genetic inheritance. Positional effects in the mammalian uterus

Hormones and Behaviour

Hormones influence sensory perception (human visual, rat odor, preference in castrated vs. intact). Hypothalamus neurons monitor the internal state (thirst, hunger) and send neurosecretory cell axons to the posterior pituitary. In females, Oxytocin, released from the posterior pituitary, triggers milk let-down in mammary glands or contraction in the uterus during child-birth. It is also associated with powerful emotional effects (affective states) in parental behaviour and maternal competence. Moreover, the hypothalamus secretes releasing factors to stimulate or inhibit manufacture and secretion of hormones in anterior pituitary portal system: Prolactin, Follicle stimulating hormone (FSH), Luteinising hormone (LH); receives, stores and releases neurohormones from hypothalamus.

Expression of sex differences in behaviour also requires that steroids activate many aspects of the phenotype during maturation. Reproductive Neuroendocrinology of Ring Doves (Streptopelia risoria): monogamous, sexually monomorphic, parental care from both male and female (incubate eggs, crop "milk"); Lordosis in female hamsters: female soliciting behaviour. Lee-Boot effect—estrous cycles slow and stop in female mice housed together; Whitten effect—individuals start cycling again in synchrony if exposed to male odor; Vandenbergh effect—acceleration of onset of puberty in female rat when exposed to odor of male; Bruce effect—failure of recent pregnancy of female rat when exposed to male who is not the father.

Arthropod Molting

Hormonal Control of Arthropod Molting: Brain: releases ecdysiotropin which stimulates release of molting hormone (MH) from Prothoracic Gland; Corpora allata: Juvenile Hormone (JH) suppresses metamorphosis; Prothoracic Gland; Corpora allata: Juvenile Hormone (JH) suppresses metamorphosis.

BIOLOGICAL RHYTHMS

Our surrounding undergoes predictable, cyclical fluctations as a result of changes in seasons, or the time of day. When events and conditions repeat in a rhythmic pattern, the ability to predict, anticipate, and prepare for them is a highly beneficial trait. Not surprisingly, most animals are able to adjust their physiology, behaviour, and life cycles to the upcoming conditions. Owls begin to stir in the late afternoon and evening, awaiting the emergence of small mammals at dusk. Ground squirrels gather rations and pack on fat reserves in the fall in preparation for cold winters spent underground. Moose reproductive cycles match the birth of fawns in the spring to the rich emergence of forage at that time.

Human core body temperature cycles with a low during the middle of their sleep cycle and highs around lunch time and early evening. Bioluminescent fireworm of Bermuda time their romantic gettogethers to diurnal as well as lunar cycles. Adult amergence of 17-year cicadas, or bamboo flowering every few decades, is closely timed across entire populations. Chronobiology, the science that describes timing in biological clocks and their associated rhythms, makes extensive use of terms and concepts derived from engineering disciplines. Rhythmic events are described as to their period (i.e., the amount of time it takes to go from peak to peak), frequency (i.e., the number of cycles completed within a specific unit of time), amplitude (i.e., the distance between peaks and average), and phase (i.e., the timing of the rhythm relative to some objective, external point in time).

The timing of a cycle's events can operate at many different time frames. Some cycles are link to obvious external patterns including:

— sun: circadian refers to 24h period length, ultradian to cycles exceeding 24h

— moon: lunar cycles, lasting 24.8h, are controlled by the phase of the moon, while tidal rhythms of 12.4h are determined by the timing of its gravitation pull

— tilt of earth's spin axis: a circannual rhythm spans events that repeat on a 12 month basis corresponding to the earth's seasons.

Biological Clocks

Many observed rhythms in physiology and behaviour often crucially depend on the presence of endogenous cycles and their production through biological clocks. Periodic rhythms, which are not simply responses to external periodic cues, have been documented for most living beings, including bacteria, fungi, plants, and animals. Biological clocks are self sustaining oscillators which will continue a period of free-running cycling even in the absence of external cues. However, clocks are usually linked to and can be reset by the environment via cues (i.e., Zeitgeber). Such entrainment keeps an organisms clock synched to its surrounding conditions.

Circadian clocks must contain a minimum of three basic elements:

(1) input pathway(s) relay environmental information to a circadian pacemaker;

(2) the endogenous pacemaker (oscillator) generates temporal patterns; and

(3) output pathway(s) for the pacemaker to regulate output rhythms.

A search for biological clocks commonly attempts to isolate an organism from its external cues and to search for the continuation of rhythmic patterns. Isolation from all possible (e.g., geophysical, magnetic, or radiation) cues remains a difficult task, however, animals will display cyclical activity patterns even when maintained in constant conditions aboard a spacecraft orbiting far above the earth.

Genetic components of the biological clock have emerged from research on the fruit fly where mutant lines of flies displayed abnormal cycling, including a shorter period, a longer one and its complete absence in yet another. All three mutations were mapped to the same gene—termed period or per. Disruption of the same gene in human sleep disorder underscored the conserved nature of the molecular circadian clock throughout evolution.

Biological Clocks exhibit a high degree of inheritance, independence of temperature and social conditions, strong resistance to pharmacologcial and chemical disruption, and may even be expressed at the level of single cells. Entrainment is often limited to a narrow range of possible outcomes, may utilise a small set of cues, and its timing is critical.

Human circadian rhythms free-run at slightly longer than 24 hours and are reset to the normal 24h day by light cues. Humans are able to readily adjust to a 23-25h day but not to a 22 or 28h day. Biological rhythms may be expressed over many different time frames, and entrainment of one clock may disrupt multiple rhythmic changes suggesting a basic linkage of patterns (e.g., circadian activity and estrous cycles in mammals).

Karl von Frisch's search for biological clocks characterised a periodic, night-time lightening of skin chrmatophores in minnows. These cyclical changes in skin pigments persisted even in blinded individuals. With damage to the pineal gland, however, fish were no longer able to change skin color in rhythmic fashion. The existence of extra ocular light perception in the brain's pineal gland allowed the fish to change color even when ocular light perception was disrupted.

Molecular Basis of Cellular Clocks

The presence of rythmic patterns even at the level of single cells indicates the presence of cell-autonomous circadian mechanisms. In the fruit fly (Drosophila) two proteins (PER and TIM) play a role in rhythmic patterns at the cellular level. The two proteins form a dimer complex in the nucleu, binds to the gene's promoter, and inhibits production of further per and tim RNA.

With gene transcription of the component proteins turned way down, few new PER and TIM proteins are made. Existng PER-TIM complexes are gradually degraded, and decrease in numbers as they are not replaced. The protein-induced inhibition of per and tim transcription weakens, mRNAs for them increase, leading to enhanced PER and TIM protein, complex formation between them, and renewed inhibition of per and tim gene transcription.

In the mammalian suprachiasmatic nucleus entrainment derives from light-induced expression of transcription factors (ie., immediate early genes, IEG), which control the expression of timing genes down-stream. At the output level, the circadian clock controls the transcription of a number of genes such as CREM (i.e., CRE modulator).

Suprachiasmatic Nucleus

The mammalian circadian system relies on oscillating neurons grouped within the suprachiasmatic nucleus (SCN), a distinct group of cells located in the hypothalamus. Its destruction results in the complete obliteration of regular sleep/wake rhythms. Cultured SCN cells maintain their own rhythm in the absence of external cues. The SCN receives information on day length from the retina, interprets it, and during the dark phase enhances secretion of the hormone melatonin from the pineal gland.

Sleep

Sleep, an essential state of natural rest, is observed in most animals. It is characterised by reduced voluntary body movement, decreased reaction to external stimuli, and a loss of consciousness. Humans sleep an average of 7.5 hours, during which they proceed in around 100 minute cycles of two broad types. Rapid eye movement (REM) occupies roughly 25% of sleep time. During this stage the body shows a loss of skeletal muscle tone while the brain is quite active.

Electroencephalography (EEG) shows a mixture of frequencies that is similar in appearance to the wakeful EEG and coincides with periods of intense dreaming. Non-rapid eye movement (NREM) sleep occupies most of the sleep time and is charcterised by frequent limb movements and sleep walking while the brain exhibits reduced activity and little dreaming.

The inherent function of sleep is not yet completely understood. Sleep, which affects the immune system and metabolism, supports a restorative function with increases in wound healing and the production of anabolic growth hormones. Sleeps is marked by a variety of physiological processes of growth and rejuvenation of the organism's immune, nervous, muscular, and skeletal systems. In newborns sleep appears to play a critical role in establishing proper functional connectivity in the brain. A number of studies suggest the existence of a correlation between sleep and the many complex functions of memory. Memory and learning involve nerve cell dendrites sending information to the cell body to be organised into new neuronal connections. Sleep which reprocesses the days events in the absence of external information is processed by these dendrites as memories are solidified and knowledge is organised.

REFERENCES

Hailman JP. 1977. *Optical Signals: Animal Communication and Light.* Indiana Univ Press, Bloomington.

Konishi M. 1989. "Birdsong for Neurobiologists". *Neuron* 3: 541-549.

Kuhl P. 2000, *A new view of language acquisition*. PNAS 97(22): 11850–11857.

Lapidge KL, Oldroyd BP, Spivak M. 2002. "Seven suggestive quantitative trait loci influence hygienic behavior of honey bees". *Naturwissenschaften* 89(12): 565-568.

Plomin R, Fulker DW, Corley R, DeFries JC. 1997. "Nature, nurture and cognitive development from 1 to 16 years: A parent-offspring adoption study". *Psychological Science*, 8, 442-447.

Robinson GE. 2002. "Genomics and integrative analyses of division of labor in honey bee colonies". *American Naturalist* 160: S160-S172.

Roeder KD, Treat AE. 1957. "Ultrasonic reception by the tympanic organs of noctuid moths". *J. Exp. Zool.* 134 : 127-158.

Rothenbuhler WC. 1964. "Behaviour genetics of nest cleaning in honey bees. IV. Responses of F1 and backcross generations to disease-killed brood". *Am Zool* 4:111–123.

Roberts, D. 1996. "Facilitating Transformative Behaviors in Animals." *Animal Education Quarterly* (Vol. 47, No. 1). PETS Social Science Publishing.

Williams GC. 1966, *Adaptation and natural selection: a critique of some current evolutionary thought*. Princeton, N.J: Princeton University Press.

8

Animal Social Behaviour

Animal behaviour is the bridge between the molecular and physiological aspects of biology and the ecological. Behaviour is the link between organisms and environment and between the nervous system, and the ecosystem. Behaviour is one of the most important properties of animal life. Behaviour plays a critical role in biological adaptations. Behaviour is how we humans define our own lives. Behaviour is that part of an organism by which it interacts with its environment. Behaviour is as much a part of an organisms as its coat, wings etc. The beauty of an animal includes its behavioural attributes.

For the same reasons that we study the universe and subatomic particles there is intrinsic interest in the study of animals. In view of the amount of time that television devotes to animal films and the amount of money that people spend on nature books there is much more public interest in animal behaviour than in neutrons and neurons. If human curiosity drives research, then animal behaviour should be near the top of our priorities.

Research on animal behaviour and behavioural ecology has been burgeoning in recent years despite below inflation increases (and often decreases) in research funding. Two of our journals Animal Behaviour and Behaviour Ecology and Sociobiology rank in the top six behavioural science and zoological journals in terms of impact as measured by the Science Citation Index. From 1985 to 1990 Animal Behaviour has grown from quarterly to monthly publication and its page budget has more than doubled. Many related journals have increased their size and frequency of publication in the same period. Ours is an active and vital field.

While the study of animal behaviour is important as a scientific field on its own, our science has made important contributions to other disciplines with applications to the study of human behaviour, to the neurosciences, to the environment and resource management, to the study of animal welfare and to the education of future generations of scientists.

NEUROETHOLOGY

Sir Charles Sherrington, an early Nobel Prize winner, developed a model for the structure and function of the nervous system based only on close behavioural observation and deduction. Seventy years of subsequent neurobiological research has completely supported the inferences Sherrington made from behavioural observation.

Neuroethology, the integration of animal behaviour and the neurosciences, provides important frameworks for hypothesising neural mechanisms. Careful behavioural data allow neurobiologists to narrow the scope of their studies and to focus on relevant input stimuli and attend to relevant responses. In many case the use of species specific natural stimuli has led to new insights about neural structure and function that contrast with results obtained using non-relevant stimuli.

Recent work in animal behaviour has demonstrated a downward influence of behaviour and social organisation on physiological and cellular processes. Variations in social environment can inhibit or stimulate ovulation, produce menstrual synchrony, induce miscarriages and so on. Other animal studies show that the quality of the social and behavioural environment have a direct effect on immune system functioning. Researchers in physiology and immunology need to be guided by these behavioural and social influences to properly control their own studies.

SOCIAL ETHOLOGY

Lorenz and Tinbergen usually studied solitary animals (*monads*) and pairs of animals (*dyads*). Starting in the 1960s, researchers started to put more emphasis on larger groups of animals. Crook (1970) proposed the term *social ethology* to describe this emphasis in animal research. The new ethologists focused on competition, communication, and cooperation between animals—how they acquired and maintained territories, maintained appropriate population levels, harvested food, defended themselves, regulated combat, and formed cooperative relationships.

Animals who live in groups develop distinctive ways of interacting. Social ethology commonly involves *predator/prey detection* (tracking, avoiding), *alarm responses* (signals of danger shared with other prey animals), *competition for status* (threat displays, ritual combat, submission postures, tournaments, status hierarchies) and *prosocial behaviour* (e.g. invitation displays, grooming, and play) plus the *reproductive cycle*—attracting a mate, mating, and raising offspring.

SPACE UTILISATION

Individuals must find their place within the three dimensional expanse in which all material objects are located. Individual species ought to strive for finding the most favorable combination of spatial distribution and the use of resources offered there. Distribution and behaviour of individuals or species contingent on spatially explicit references.Habitat refers to any part of the biosphere where a species can successfully live and reproduce; Niche, in contrast, describes a species' ecological role within a community.

Individuals

— *Ideal Free Distribution.* The relative number of foragers in each patch should match the relative, available amount of resources obtained there. Individuals should distribute themselves among habitats so that every individual maximises its net rate of return. Costs and benefits may depend on how many other users are already using the resource. applies to food, mates, etc.

— *Ideal Depotic Distribution.* Individuals chose the most suitable area, within the constraints that other residents provide

— *Allee-type Distribution.* Individuals chose the most suitable area, within the constraints that other residents provide

A variety of factors may determine the presence or absence of a species at a particular site. Individuals pf a species may not have reached the habitat - dispersal. Alternatively, a variety of determinants may play a role, including abiotic (e.g., weather conditions, soil, climate) or biotic (interactions with other organisms, predation, competition, disease, social factors).

Habitat Preferences

Natural selection will favor individuals which utilise those habitats in which the greatest number of successful offspring can be raised. Spatial constraints may involve a balance of considerations: e.g., Herring gulls; Variables to consider: resource limitations; spatial and temporal distribution of the resource; variation in resource quality; number of competitors that attempt to control a resource; Predation pressure, etc.

— *Habitat preference as a closed developmental system:* many invertebrates, crayfish. Host plant preference in bladder grasshoppers. Poor response to rapid environmental change and human impact, e.g. Karner blue butterfly (Lycaeides melissa samuelis) depends on lupine (Lupinus perennis) as its sole larval food source. Species with highly fixed, genetically programmed habitat selection may require considerable time to evolve mechanisms for the selection of a new habitat

— *Habitat preference as an open developmental system:* habitat imprinting: many migratory vertebrates, such as salmon tradition: mountain sheep, greylag goose migration

— Dispersal Ecological process affecting distribution, and genetic process affecting geographic differentiation and variation; leaving an area of birth or activity; largely a short range effort; Colonisation - long range dispersal. Stay in the safety of home: little danger but competition may reduce reproduction or Disperse and take chances: high mortality but high payoffs. Natural dispersal: e.g. Cattle egrets or Intended and unintended aid in dispersal by humans: e.g. European starling, house sparrow, Africanised honeybees

— *Descriptives for spatial distributions*: Are distributions random, clumped, dispersed? Describe distribution with: minimum polygon, bivariate normal, spatial autocorrelation, contiguous clustering, or non-parametric utilisation distribution (UD).

Predator-Prey Arms Races

The constant competition between predators and prey animals is a major stimulus to evolution, sometimes called an evolutionary "arms race." Predators adapt to prey populations, and the prey populations adapt to the innovations of a predator.

Bats have a very effective system for locating insects, using a high-frequency sonar that bounces off insects and provides directional information. Bats are also excellent flyers capable of turning on a dime. Insects that cannot escape the bats are more likely to be eaten. Insects that develop a genetic mutation reducing the chance of bat predation are more likely to survive and reproduce. For example, the Noctuid moth has a simple three-neuron auditory system tuned to the frequency of the bat's sonar. When the neurons fire impulses (indicating that a bat is nearby) this triggers a motor program that causes the moth to dive into the grass, where it is safe from bats.

The praying mantis-itself a predator for many smaller insects-is just prey to bats. So the praying mantis evolved a special ear in the middle of its thorax, tuned to the exact frequency of the sound emitted by bats. When stimulated, the receptor triggers a "sudden full extension of its forelegs" and contraction of the abdomen leading to "an abrupt and dramatic deviation in its flight path". This helps the mantis avoid being eaten.

As prey species evolve successful defenses, predator species evolve better techniques for hunting. Some bats employ unusual echo frequencies, so they overcome specialised defense systems tuned to the usual frequency of bat sonar. Some bats use non-echo hunting strategies, so they can catch insects that have defenses only against the echolocation strategy. The predator-prey competition stimulates innovation on both sides.

Animals commonly have built-in circuits to trigger avoidance of stimuli indicating danger. For example, a Central American bird called the motmot preys on snakes, but it must avoid the poisonous coral snake. Coral snakes have a distinctive color pattern consisting of red, yellow and black rings, so motmots are born with a fear of this pattern.

Eye Spots

Throughout the animal kingdom, insects preyed upon by birds often have patterns of pigment that resemble eyes. Eyespots scare birds away or divert their pecks to the wing instead of the body. The secret to understanding how tiny eyespots could scare birds is to realise that birds are *pattern-recognising* experts, but they are *relatively insensitive to the size of a pattern*. They have to ignore size because they fly at different altitudes and must recognise visual patterns no matter how large or small they might appear from various heights. So birds respond to the *configuration* of a pattern more than its *size*. If an insect preyed upon by birds happens to develop a pattern that resembles an eye spot, it is less likely to be eaten and more likely to survive and reproduce.

Why are birds instinctively afraid of eyespots? Predators, as a rule, have large front-facing eyes. This helps them localise their prey exactly, using binocular disparity. Animals that prey on birds—owls, primates, and cats—typically have large, staring eyes. Therefore the birds are showing a biologically prepared form of avoidance. Birds that avoid eye-like patterns are themselves more likely to survive and reproduce.

There is a second way in which eyespots protect butterflies. Birds that attack the insect often peck at the eyespots, leaving the butterfly or moth free to fly away. Swynnerton painted conspicuous eyespots on butterflies visiting clumps of bananas next to his verandah. He caught the painted butterflies later and observed that 47 of 51 butterflies had injuries from bird pecks. Of the 47 injuries, 36 were at or near the eyespots. A peck on the eyespot may reduce the butterfly's flying efficiency, but at least the butterfly can survive to reproduce.

Figure 1. Eyespots protect butterflies

Eyespots are so common among butterflies and moths that they have attracted considerable research attention. A team at the University of Wisconsin identified specific genes and proteins that controlled eyespot formation. Within a few years, they identified the specific sequence of changes in gene expression that led to the evolution of eyespots.

Visual Mimicry

Sandved came across a butterfly in Thailand with a repeated pattern of pigment that bears a remarkable resemblance to a small monkey face. It is only about as big as a pencil eraser, and it is repeated along the edge of the butterfly wing. Monkeys are predators of birds in Thailand, so this might be an effective way to scare off birds that might otherwise eat the butterfly.

Presumably the first occurrence of the pattern was much cruder than this, a lucky accident that helped the chrysalis avoid being eaten by birds. With succeeding generations, the pattern was refined, as chrysali with a cruder (less face-like) patterns were eaten by birds. The more face-like the pattern, the more likely the butterfly was to survive and reproduce. Eventually the artwork became quite accurate.

This example is so dramatic that it could easily provoke skepticism. After all, it is a sketch, not a photograph. Might the unknown artist have taken some liberties? However, the *spalgis epius* has the nickname *apefly*, so the resemblance has been known and remarked-upon for a long time.

Figure 2. Chrysalis of the spalgis epius butterfly

Earlier we made the point that evolution favors any adaptation that encourages successful reproduction of the species. We also made the point that birds respond to highly specific visual stimuli of evolutionary significance. Now we see how these factors come together. One species (the potential prey) evolves a visual pattern (eye spots or a monkey face) that acts as a significant sign stimulus triggering avoidance in other species (the bird predator).

These are all examples of visual mimicry or imitation. Humans are highly visual animals, like birds. Like birds, we cannot help but see a face in a few blotches of pigment on the spalgis epius chrysalis. Our brains respond to significant visual patterns automatically. Other species respond more to mimicry of odors, textures, and sounds. Odor and sound mimicry are also widespread in nature.

Alarm Responses

Probably the best way to avoid being eaten by a predator is simply to stay out of its way. The slap of the beaver's tail, the flash of a white tail-patch of a deer, the "chip chip" of a bird, the "scolding" of a squirrel are all *alarm responses*. They warn of danger in the presence of a predator, allowing prey animals to flee.

Vervet monkeys have at least *three distinct alarm calls* for three different predators: leopards, eagles and snakes. The alarm signals lead to distinct behavioural responses. When Seyfarth and colleagues played tape recordings of alarm calls to a group of free-ranging (non-captive) vervets, the leopard-type alarm call made the monkeys run into trees. The eagle-type alarm call made them look up at the sky. The snake-type alarm call made them look down to the ground.

This finding led to a flurry of research documenting variable and informative aspects of alarm calls in many species. For example, the little black-capped chickadee, a bird commonly found in American neighborhoods, varies its alarm call depending on the predator, and its alarm call is understood by other birds such as nuthatches.

When a chickadee sees a predator, it issues warning call-a soft "seet" for a flying hawk, owl or falcon, or a loud "chick-a-dee-dee-dee" for a perched predator.

The "chick-a-dee" call can have 10 to 15 "dees" at the end and varies in sound to encode information on the type of predator.

Alarm responses are often *shared by several species*. This benefits every animal that participates in the shared alarm system (usually a group of prey animals). The result is an alarm system more sensitive than any one species could provide on its own. Watering holes on the African savanna are frequented by all sorts of prey animals, such as baboons, birds, and cattle. All are threatened by the area's chief predator: the lion. When a member of any prey species senses the approach of a lion, it makes an alarm call and all the animals of all the various prey species flee.

Humans participate in a variety of shared alarm calls. A watchdog shares its alarm call with its owner when it barks at shadowy objects in the night. Pet birds and cats also warn their owners of intruders or dangers, on occasion. Burden collected a "bulging file" of newspaper clippings about cat heroes.

Take a family in Lewiston, Idaho, whose mistress decided to take a nap in the middle of the day. About two hours later her pet cat Tiger roused her with loud and anxious mewing. Following Tiger into the spare bedroom, the lady found pieces of ceiling falling onto the floor in flames. She managed to rescue Tiger and her kittens, but it was only a matter of minutes before the whole house went up in flames.

The human alarm call is the *scream*. It emerges in stereotyped form, without learning, in normal members of our species. Small children commonly scream as an involuntary response to certain stimulus situations, such as being approached by an adult playing the role of a monster. Like other species' alarm responses, a scream can be detected at a large distance, providing a useful warning.

Animal Movement

Predicatable fluctuations in environmental conditions impact the behaviour and movement patterns of many animals. When nothern temperate ponds freeze over in winter, bald eagles are cutoff from their prey, forced to travel in search of open water. Animals respond by travelling to such seasonal changes in carrying capacity by seeking out abundant resources in the summer and emmigrating from limited, harsh resource conditions in the winter. Migration refers to these regular, seasonal journeys of animals that alternate between breeding and non-breeding locations. Many species are able to move great distances and excell at an ability to find their way. A combination of cues and mechanisms are employed to complete some of the most amazing achievements.

A series of independent questions can be addressed to explore the causation of migratory movements in animals, such as song birds.

— *Proximate Causation*: hormonal influences on timing (restlessness), cues for navigation.

— *Ultimate Causation*: For functional explanations of songbird migration, migrants must exceed local residents in the number of offspring, despite making up for additional energy costs and dangers associated with migration.

— *Phylogeny*: Many species of birds migrate. Fossil evidence suggests that early songbirds evolved in the tropics. Migration may have arisen when individuals moved north (or southward) in spring to exploit longer daylight hours, local resource abundance in temperate zones, and release from competition in the tropics during the breeding season with its high food demands. The precise course of current major flyways have probably evolved since the last glaciation.

— *Ontogeny*: learning of migration paths and innate preference for migration direction (European warblers, Starlings)

Homing

Homing refers to an individual's ability to return from a distant location to the only place it is certain to know something about. Animals may utilise a variety of cues and mechanisms to determine the location of its home site.

Limpets fit their shell to one specific spot on a rock. A close fit is critical to withstand the impact of waves when submerged and prevent dessication when exposed. Scraping algae from the rock surface, they travel from their home site, leaving behind a chemical trail. By reversing direction on this trail they are able to return to the very site they came from.

The homing pigeon has been selectively bred to be able to return to its home over extremely long distances. As a pigeon generally returns to its own nest and its own mate, human handlers have selectively bred subsets of birds that were particularly good at this task. Homing flights from as far as 1000 kilometers have been recorded by exceptional birds at average flying speeds of around 50km/h. For thousands of years humans have used homing pigeons to return messages back to home base by attaching them to the birds legs.

Migration

Migration includes all movements made in response to changes in food availability, habitat or weather.

Nomadic Migration

'Seasonal Migration': Travel between a summer area for breeding and an area for overwintering are driving the seasonal migration of many animals, such as song birds. The energetic advantages of breeding in a site of seasonally high food resources and extended daylight hours allow diurnal birds to produce larger clutches than related non-migratory species that remain in the tropics year-round. In autumn, as the days shorten and food sources dwindle, birds return to warmer regions where the available food supply varies little with the season. The success of migrants indicates that these advantages greatly offset the high stress, energetic costs, and other risks of the migration. Cardinals are common winter residents in northern latitudes. As members of a taxon with mostly tropical distribution, they may have reduced their winter return migration.

Return Migration

Many individuals show a distinct preference for departing in one particular direction. This applies to diurnal movements of plankton that feed at the surface during the veening and night, and sink to lower, darker regions to avoid falling prey to visual predators.

Navigation

Navigation refers to the ability of planning and controlling one's movement from one place to another. It requires knowing one's current location on earth, the location of the desired goal, and an ability to calculate travel directions to get from here to there.

Dead Reckoning

Many birds determine a particular flight direction based on environmental cues that serve as landmarks on the Earth's surface, including longitudinally magnetic lines, the lcoation and direction of the sun and other celestial objects, or prevailing wind direction and odors.

In vector navigation individuals arrive at a given destination by maintaining a compass direction (or directions) for a predetermined amount of time or distance. Dead reckoning is the process of estimating one's present position by projecting course and speed from a known past position. Honey bees for instance communicate the location of profitable nectora sources to hive mates. Using a waggle dance they are able to transmit codes for direction and distance.

European Starlings from much of continental Europe overwinter on the channels coast of France and Britain. A group of banded individuals was moved from one part of Europe to a more western geographic location. This transfer would now require a ninety degree change in direction to reach their traditional wintering grounds. Adult starlings were admirably able to perform that course adjustment and returned at their normal wintering range. Newly-hatched juveniles, however, which had never made the trip before, continued to fly their traditional southwesterly direction and ended up on the coast of Spain.

Orientation Cues

Niko Tinbergen's work on Digger wasps illustrates the importance of local visual cues to pinpoint the exact location of the burrow.

Radar images show that migrating bird flocks drift off courses in a strong crosswind, except for flocks that travel along distinct landmarks, such as rivers, coast lines, or mountain ridges. Landmarks may also arise from specific atmospheric conditions and distinct geographic features, such as Point Pelee, which jutts into Lake Erie and serves as key staging ground for migrating hawks in search of updrafts.

Visual Orientation refers to the use of geographic landmarks (rivers, coastlines, ridges, etc.). Diurnal migrants often follow such landmarks, generally of lesser importance for nocturnal migrants, although there are documented cases of nocturnal migrants following rivers or coastlines. This may be particularly important for navigation to a precise breeding or wintering locality.

TERRITORIALITY

Territoriality is a type of intraspecific or interespecific competition that results from the behavioural exclusion of others from a specific space that is defended as territory. This well-defined behaviour is exhibited through songs and calls, intimidation behaviour, attack and chase, and marking with scents. This form of defense proves to be very costly for animals. So one is forced to ask, Why do animals take part in such interspecific competition?

In order to understand this question one must take a cost-benefit approach to territoriality. The proximate reasons for such defense vary. For some animals the reason for participating in such elaborate protective behaviour is to acquire and protect food sources, nesting sites, mating areas, or to simply attract a mate. The ultimate cause of this behaviour may be attributed to the increased probability of survival and reproductive successes. In defending a territory an animal is ultimately securing that it will have an habitat in which to forage for food and to successfully reproduce, thus increasing the animal's overall fitness. This ultimate theory is strengthened when one considers the instances in which territoriality increases; in times of depleted resources the presence of territoriality increases. The presence of territoriality often forces less fit animals to live in sub optimal habitats, thus reducing their reproductive success.

Though territoriality offers immense reproductive and nutritional benefits, it also comes at a cost. Defending territory is not easy. Territoriality cost time and energy and can often interfere with other fundamental activities as parenting, feeding, courting, and mating. For these reasons territoriality may not be seen as a benefit in all animals. Animals must be able to reap the fruits of territoriality, while expending the least amount of energy. For these reasons if resources are abundant and predictable it would be disadvantageous to defend the territory. On the other hand, if resources are scarce and undependable it would be advantageous to exhibit territoriality.

An animal chooses its territory by deciding what part of its home range it would like to defend. In selecting a territory the size and the quality play a crucial role in determining an animals habitat. Territory size generally tends to be no larger than the organism than requires to survive, because with an increase in territory comes an increased in energy expenditure. For some animals the territory size is not the most important aspect of territoriality, but rather the quality of defended territory. The quality is considered to be fundamentally important due to amount of food availability and superior nesting sights. Animals depend on these features to ensure their superior fitness.

Animals invest a lot of time and energy in defending their territories, and for this reason they fight vigorously to defend their territory at all cost. Researchers suggest for this reason that when a rival challenges a territory holder, the owner almost always wins the contest. This phenomenon could be attributed to an evolutionary stable strategy which assets that rules for behaviour is controlled by an inherited proximate mechanism such that the differences between individuals in their strategies are liked to differences in their genes. Territory plays an important role as a mechanism of population regulation, insuring the success of fit animals, and aiding in the eradication of less fit animals. Territorially also plays a fundamental role as an indicator of carrying capacity; it also serves as an indicator of how much habitat is necessary to support viable populations. For these reasons researchers continue to examine the well-developed behaviour of territoriality.

Struggles for Territory and Social Status

Animals commonly stake out a territory and defend it. Studies of *nest spacing* and *population control* are related to territoriality. Some birds will nest differently depending on available space and population density. Rodents will reproduce in different numbers depending on available resources.

Even ants have territorial battles. One species in the Southwest United States has "tournaments" during which different ant colonies meet at the boundaries of their territories. They engage in mock combat, drumming on each other's abdomens with their antennae. After 10 to 30 seconds, one or the other ant yields ground and the encounter ends. In this way, territorial boundaries are re-established with a minimum of actual fighting.

Group-living animals commonly joust for *dominance*. Different animals test each other to see who is "really boss." The outcome has important implications for each animal's genetic success. Dominant animals usually reproduce more than non-dominant animals. Aggression in primates is related to the male hormone testosterone . The more testosterone a male primate has, the more aggressive it tends to be. Defeated males lose testosterone.

Similar phenomena occur in other mammals. A dominant ram mates repeatedly with his pick of the ewes. The male that loses a head-butting contest might not reproduce at all.

Aggressive interactions between porpoises, usually during dominance disputes, include striking, raking with the teeth, and ramming with the beak or rostrum, sometimes with serious consequences, such as broken ribs or vertebrae, or punctured lungs, in the rammed animal. A dolphin that has become accustomed to humans shows no hesitance in challenging the human for dominance, by means of threat displays and blows; a person who is in the water with an aggressive porpoise is at a dangerous disadvantage. The sentimental view that these animals are harmless stems at least in part from the fact that they are usually in the water and we are usually on boats or dry land; they can't get at us.

Sarah Boysen, a chimpanzee expert, made a similar point. She said chimpanzees "test" human lab assistants with dominance displays such as roughhousing and hair pulling. Chimps also seem to be sensitive to gender differences in humans. Boysen noticed that the chimpanzees seldom bother male lab assistants who are physically large, especially if they have facial hair such as beards. But the chimpanzees challenge female lab assistants "constantly, every day."

SOCIAL BEHAVIOUR OF ARMY ANTS

Army ant colony castes consist of queen ants, soldier ants, and multi-purpose worker ants. In the army ant colony, the queen ant lays eggs, the soldiers focus their attention to the defense of the colony, and the worker ants are split between being foragers or tending to the queens brood.

The Army ant is abundantly found in humid lowland forests or heavily forested areas of Central/South America. The changes in temperature and precipitation have no affect on the life cycles or migration of these ants.

The social structure and life cycles of army ants have two distinct phases. Army ant colonies go through a stationary phase and a migratory (nomadic) phase. The phases are based on the ability of a queen to produce many eggs in a very short period of time (from 100 000 to 300 000 eggs in five to ten days); thus, the queen overpopulates a given area, which then triggers migration of the colony.

In the stationary phase, the army ant queen's abdomen swells to hold 55 000 to 66 000 eggs. As the eggs develop into mature ants, tens of thousands of adult workers appear, triggering the nomadic phase: colony activity increases exponentially, and "swarm raids" increase in size and intensity. With a colony of this size, the army ant colony must emigrate daily. The migration continues until "larval pupation" begins. Once this occurs, the colony re-enters the stationary phase, and the cycle repeats.

Due to their large colony sizes, army ants migrate in order to find food. Most ants are "column feeders." Army ants are what are called "swarm feeders." The foraging workers form a fan-shaped swarm with a broad front. Along with feeding, army ants also have a unique way of forming nests. The army ant nest is made up of army ants themselves. The ants form walls and fasten onto each other by using their mandibles. The ant colony forms these walls almost anywhere in order to enclose the queen and her brood. The nest is very structured, as it provides corridors within itself for transporting food and eggs throughout the many areas of the nest.

Army ant colonies march and nest in different phases as well. In the nomadic phase, army ants march at night and stop to camp during the day. The colony begins a nomadic phase when available food has decreased. During this phase, the colony makes temporary nests that are changed everyday. Each of these nomadic rampages or marches lasts for approximately 17 days.

Army ants are notorious for eating anything that gets in their path. They can kill and eat up to 100 000 animals in a day! They normally consume lizards, snakes, chickens, pigs, goats, scorpions, tarantulas, beetles and other ants. Army ants can also climb trees, and feed on animals within its canopy.

Army ants communication is completely dependent on chemical messaging and trail pheromones. These methods of communication act as a stimuli for changing behaviour patterns. Unlike other ants, army ants do not have compound eyes, but instead have single eyes. Army ants are also blind and have to use their antennae to sense smell and touch. The army ants use these senses to communicate in nesting and raiding.

HONEYBEE FORAGING BEHAVIOURAL ANALYSIS

The common honeybee (Apis mellifera) forages around the hive in search of resources. Usually this resource constitutes food in the form of pollinating flowers. Several landmark studies were performed in the middle of the 20th century to analyse how these social organisms communicated exact distances and direction to one another in order to effectively locate these resources. Karl von Frisch determined that honeybees perform two distinct dance routines that coincide with two different distance approximations made by the foraging bee. These two dances, the Round dance and the Waggle dance, communicate to the other the approximate distance from the hive to the new resource. Only the Waggle dance communicates direction.

The Round dance is performed by the returning bee usually in complete darkness, vertically on a honeycomb. The circuitous motion attracts other foragers, which then learn that the resource is within approximately 50 meters of the hive. No direction is given by this routine (von Frisch). As a result, the newest foragers leave to search in all directions surrounding the hive. Behaviourally, this dance is energetically favorable due to the short distances travelled. In contrast, the Waggle dance is energetically unfavorable to the individual, but beneficial to the hive.

The Waggle dance is performed primarily when the resource is further than 50 meters. The returning forager either performs the dance on a vertical surface or a horizontal one. To determine distance and direction, the bee orientates itself relative to the sun. Any deviation from this point gives the angle the new foragers should pursue. If vertical, the bee orientates itself to gravity. Perpendicular to the ground becomes the reference point (i.e. the sun). Deviations from such relay direction accordingly. Distance is communicated by the length of the abdomen shake that forms the middle of a figure eight dance.

Kirchner et al. determined that the Round dance does not convey direction. By tipping several hives onto their horizontal axis and placed a resource 10 meters away, the reference heading was experimentally altered. Regardless of the hive orientation, the number of successful returning foragers was constant. Once the distance crossed 50 meters, the success rate decreased for hives orientated horizontally, indicating direction is dependent on hive orientation for the waggle dance to be effective.

Honeybees foraging behaviour can be interpreted from a historical or adaptationist view. Lindauer performed historical analysis that reveals that distantly related bees might have evolved through 3 stages of development: First stage analysis reveals that the genus Trigona conveyed direction by buzzing to gain their hivemates attention. The odors trapped on the bee trigger others to forage in search of the odor. The secondary stage involves marking the path from the resource to the hive with mandibular pheromones. She then buzzes, and the bees follow the scent markers. The third stage involves an in-flight "waggle dance" in the direction of the resource.

Seely et al. provides an adaptationist theory on bee dance evolution. The foragers seek new resources to provide increased hive fitness. By dancing, the cost/benfit ratio is reduced to the individual level more so than a group of foragers expending energy searching in all directions all the time. Their study demonstrates that dancing has become an evolutionarily desirable trait because less time foraging allows for more time collecting the resource. Honeybee dancing provides an evolutionarily advantageous behaviour which optimises the hive's fitness.

PENGUIN LOVING

Emperor penguins, Aptenodytes fôrsteri, are the largest of the penguin family. They live in harsh environments of the Antarctic, battered by extreme weather conditions and vicious predators, and are the only warm-blooded animal to winter on the open ice. It is therefore no surprise that these birds present unique behaviours. The mating and brooding habits of these penguins reflect how these penguins have adapted to overcome these formidable obstacles.

Unlike other cohabitants of the Antarctic Circle, emperor penguins mate in the winter months of April and May. Males grow yellow tufts in the areas surrounding their ears, suspected of being an attraction factor for females. Unlike other species' males that compete for nesting space and attract the females through quaint rituals (e.g. Adele penguin males throw scarcely available rocks at the feet of females to "get their attention"), emperor penguins do not create nest and put on mating displays that last for weeks, during which the future parents learn each other's call. After copulation and, about 15 days later, the laying of a single, soft-ball sized egg, the female leaves to the sea where she embarks on a mission to feed well. Meanwhile, the father cares for the egg. To protect it from the frosty floor, for the next 6-8 weeks the father balances the egg on its two feet, under the protective blubbery flaps of its underbelly, to keep it warm. During the whole incubation period the father does not eat and may lose a third to half of its body weight.

The mother returns right before hatching (usually at the beginning of September), well-fed and with food for her offspring. The two parents apparently find themselves through distinguishable calls. Visual aids are useless in a crowd of 6000 penguins that all look similar. The mother then feeds the chicks for the first time, regurgitating some of the spoils caught while she was away. As the female broods the young, the father is free to replenish his energy, only to return to jointly rear the chick along with the mother.

The newly hatched chicks are barely protected from the sub-zero temperatures by their downy coats. Another important behaviour, displayed similarly between adult penguins and chicks, is the formation of clustered groups called crèches. Within these groups the birds are able to maintain temperatures of about 96 degrees F, in contrast to the minus 30 degrees winds. So much heat is created in these crèches that the group must constantly move from location to avoid falling into the melted holes in the ice created by the heat generated by these crowds. This story ends when penguin chicks, huddled in groups at the edge of the sea, hop onto a recently broken-off chunk of ice and float towards the north to go on with their lives.

"How" do these behaviours occur? What are the proximate reasons of these behaviours? These breeding and brooding behaviours are influenced by many factors. Studies have shown that anatomical preparation for copulation is not stimulated by the mating displays, as is seen in other species. Penguins that reach the mating areas have seven times the normal amount of luteinising hormone (LH), testosterone (in males) and estrogen (in females) upon arrival, suggesting that geographical or environmental cues activate gonodal preparation, specifically corresponding to maximum size of the gonads.

Between copulation and egg-laying, levels of all hormones begin to fall, but LT remains slightly above average in incubating males, and in both males and females that are rearing their young. This leads to the conclusion that LT is responsible for incubating, brooding and territorial behaviour. The hormone prolactin has also been linked to parental care.

Next question should be why these behaviours happen? What are the ultimate causes? Why do penguins endure the harsh Antarctic winter and choose to have their young during this apparently dangerous time? Speculation has lead to a few hypotheses. One is the lack of predators. These penguins are the only animals left during the winter's cold. There are no leopard seals to menace the young. Another advantage of wintering the summer grounds is the first access to the abundance of food that the early warmer season brings. The quantity of food and the climate is perfect for the chicks to learn to fend for themselves. These two factors may have been enough evolutionary pressure to give rise to these vital behavioural traits in emperor penguins.

NESTING BEHAVIOUR

Over the course of a turtle's nesting season, a female sea turtle will leave her homing ground and migrate to nesting grounds to build a nest and lay eggs. One species of turtle, Chelonia mydas, the Green sea turtle, lives off the coasts of Brazil but nests each year on the beaches of Ascension Island, a tiny island in the mid-Atlantic Ocean (Papi et al.). These turtles typically make their way to the island between the months of December and March (Sims).

After migrating to a nesting ground, it is time for the female to choose a nesting site along the beach. Many factors play into this decision and a female will scope out a site for a few days before she actually decides to build a nest. However, once the decision of where to nest has been made, females will most likely nest at the same beach and site for the rest of their reproductive years (CCC&STSL).

All species of sea turtles exhibit similar nesting behaviour, possibly differing only in inter-nesting period length (the period of time between two nestings during the same nesting season). On the night of nesting, a sea turtle will emerge from the sea and scope out her surroundings one more time before digging a nest. If there seems to be any unnatural distractions (i.e. lighting or human presence), a female may return to the sea without building a nest ("false crawl").

Most often, however, she will proceed to dig a nest on the beach. To do this, the female turtle first digs a relatively shallow depression, known as a body pit, in the sand using her front flippers and body to push sand out of the way. Then she will excavate a single egg chamber, a deeper cavity opening to the body pit, into which she will oviposit an entire clutch of eggs during nesting. Finally, after constructing this egg chamber, she will begin ovipositing her eggs. It is extremely rare that a sea turtle will stop nesting before laying is complete but will happen if the female senses extreme danger.

After laying eggs, the female will camouflage her nest by filling in the body pit with sand and return to the sea only to nest at the same site at a later date in the nesting season. After nesting a second time, the female sea turtle will return to the sea and migrate back to her homing ground.

Many nesting grounds, such as that of the Green sea turtle, are extremely far from the species' homing ground. This migration of sea turtles during the nesting season has led to the development of many different hypotheses of how these adult female turtles continually find the common nesting ground and then how offspring navigate to the homing grounds. First, it has been suggested that sea turtles may use the earth's magnetic field to aid in navigation between homing grounds and nesting grounds. There have, however, been experiments that have weakened this hypothesis in adult sea turtles.

One such experiment involved placing magnets on the back of experimental Green sea turtles in order to disrupt the magnetic field possibly detected by the sea turtles and the courses these turtles followed on the return journey from Ascension Island were tracked. The results show that, although hatchlings may be able to navigate via earth's magnetic field, it is very possible that this ability is absent in adult. However, another hypothesis has been suggested and states that turtles may use olfactory cues from the ocean currents and winds to navigate through the ocean to nesting grounds.

THREAT, FEAR, AND INTENTION MOVEMENTS

Angry or aggressive-looking postures, noises and facial expressions are widespread among different animal species. Displays often serve as a substitute for actual fighting. One of the most common threat displays is the *stare*. "Animals as diverse as crabs, lizards, and birds all perceive staring as a threat."

When a larger or more dominant animal makes a threat display, a younger or more submissive animal usually backs down, and violence is averted. A low growl accompanied by a stare is an unmistakable threat display used by many large mammals. A common threat display for the dog is a low growl, with ears laid back and teeth bared. Cats have a threat display also: they stare, make a low growling sound, and adopt a posture that indicates they may attack. If seriously threatened, they hiss, arch their backs, and fluff their fur in the classic "cat fight" posture. Such a display might be considered a *fear* display as much as a threat display. Fear and threat displays are often similar.

The primate grimace known as the *threat face* tells an aggressor to "back off." Dr. Dian Fossey, whose life was portrayed in the movie *Gorillas in the Mist*, used her knowledge of social displays among gorillas.

In response to a silverback that would not stop bluff-charging her, she made a fright face, [a] kind of horrible grimace... The startled silverback sat down at once and began to eat, nervously, with one eye on her. Then he got up and walked away.

Many threat displays involve *intention* movements: movements that indicate the animal is getting ready for an action. In humans, a threatening intention movement is the clenched fist. In seagulls the threat posture is an upright head, ready to give a sharp peck. Cats, when bothered, may open their mouths and look like they are ready to bite, or they may administer a soft "mock bite" that does not break skin but conveys an unmistakable message: "Stop what you are doing." Mother cats do the same thing to kittens.

DISCIPLINE

A German Shepherd trainer who specialises in problem cases says his technique for greeting a disobedient dog is to clasp the dog's head and stare it in the eye until the dog looks away, indicating submission. That establishes who is boss. Cesar Millan, the "Dog Whisperer" of Animal Planet fame, pins a dog to the ground at the neck, the way a dominant wolf treats a subordinate wolf. Cesar uses his hands instead of his teeth, but the message is the same: I'm the boss, stop misbehaving. Dogs respond to this almost every time. Cesar's strength and calm, dominant manner also contribute to his uncanny success with problem dogs.

Not all animals respond so readily to assertions of authority. As a rule, *only group-living animals* are "programmed" to submit to dominant animals. In group-living species, submission is adaptive; it gives the subordinate animal better odds of survival and reproduction. Badgers, by contrast, are solitary animals. Therefore badgers are not programmed to respond to submit, and they do not respond well to acts of discipline by a human. Eibl-Eibesfeldt wrote:

> When I reared a badger I could never forbid it to do anything. If I scolded it when it opened a cupboard and pulled out my linen, the most it did was to stare at me, and if I gave it a smack on its nose it attacked me. It would not subordinate itself. A dog, on the other hand, quickly learns to obey. (Eibl-Eibesfeldt, 1970)

SUBMISSION AND APPEASEMENT

Submission postures are body positions that a submissive animal uses to acknowledge a dominant animal and ward off aggression. The submission conveys the message, "OK, you're bigger. I am no threat to you." Dogs (as well as humans) cringe and cower as symbols of submission. When dogs fight, the fight always ends the same way: the loser turns over on its back to signal submission. We once had a little dog (a Jack Russell terrier) that did not wait for a fight; whenever it met another dog, it flipped over on its back right away. The gesture worked; it never had to fight.

Submission postures often involve *exposing a vulnerable part of the body* such as the neck. A supplicant bowing before a king illustrates a submission posture. In the kneeling position, a person is vulnerable to being struck on the neck or back of the head. The "kowtow" of medieval China was an extreme example; visitors to the emperor literally threw themselves to the ground, touching it with their foreheads. Bowing remains a gesture of respect in many countries today, and to a Muslim, touching the forehead to the ground is a way of showing respect to Allah.

Submission postures are called *appeasement displays* if used to appease (lessen the anger of) an aggressive animal. One striking characteristic of animal and human appeasement strategies is *infantilisation*. Cowering, whining, crying, begging, and nervous laughter-all are responses that partially mimic the behaviour of children. All can make an aggressive animal decide not to attack.

PROSOCIAL (FRIENDLY) INTERACTIONS

Invitation Display

Invitation displays among group-living animals are ways of inviting friendly interaction. Long-necked birds such as ostriches commonly crane their necks upward and gather in a friendly group, as if engaging in a dance.

The photographer Marion Kaplan observed a group of ostriches apparently attracted by the upright rifles of marching soldiers.

The event took place in Tsavo National Park in Kenya, in 1969. "A group of park rangers were practicing their drill as several ostriches, wild but used to human presence, wandered near by. Quite soon the ostriches, unherded and without encouragement, were in a tidy line behind the rangers. Their mood was curious, playful, imitative, not at all threatening" (Marion Kaplan, personal communication). Kaplan adds that the rangers enjoyed the ostriches, but higher authorities disapproved, so ultimately the two groups were separated.

Grooming

Grooming is a common prosocial activity, so group-living animals commonly use a posture suitable for grooming as an invitation display. Dogs sometimes like to be scratched on their bellies, so they convert the universal canine submission posture (lying on the back) into an invitation display. They expect to be scratched by a friendly human in this position, particularly if they have a history of belly rubs.

Cats also use the lying on the back posture to indicate peaceful intentions, but (with rare exceptions) they do not use it as an invitation display. For cats, it is appeasement gesture, showing trust and affection but not inviting contact. Therefore, cats normally do not like a human touching their underbelly fur. When they are putting on a submission display to a human by rolling over on their backs, they do not want to be touched. People who own dogs sometimes misinterpret a cat's display as an invitation posture. Thinking the cat wants to be scratched, they reach down to scratch the cat. To most cats, that is unexpected, and it is a violation! The cat bites or scratches or flees.

Monkeys solicit grooming from other monkeys by presenting their backs. If the other monkey is interested in a grooming session, it will pick through the presenting monkey's back fur, removing dirt or insects. After a while the grooming monkey will turn and display its back to the other, which will groom it in turn.

Psychologist Barbara Smuts spent three years living with baboons to study their prosocial behaviour. The picture shows "Pandora" grooming "Virgil." You can see Virgil's unmistakably blissful expression. In this case, Virgil's belly fur was being groomed.

Smuts used the technique of *habituation*. Smuts spent every day with the baboon troop but refused to interact with them. Eventually the baboons simply ignored her and went about their business.

Smuts documented many examples of friendly or prosocial behaviours. She found that long-term friendships and alliances often extended to different age groups, such as parents and children. The blissful scene of Virgil and Pandora in the preceding photograph was soon transformed into a scene of play.

After a few moments they were joined by two of Pandora's offspring... Pyrrha was in a rambunctious mood, and she used Virgil's stomach as a trampoline, bouncing up and down with the voiceless laughter that accompanies baboon play. Every now and then Virgil opened his half-shut eyes, and, gently touching her with his index finger, he grunted as if to reassure her that he did not mind the rhythmic impact of her slight body against his full stomach.

CONTACT COMFORT

Harry Harlow, whose learning-to-learn experiments we reviewed in the earlier section on Comparative Psychology, set up one of the nation's best-equipped primate laboratories at the University of Wisconsin. There he did a famous series of experiments focused on *contact comfort* in babyhood.

Harlow raised rhesus monkeys from birth, and he found it was necessary to keep young monkeys separate from each other for health reasons. This separation began soon after birth, and he noticed that young monkeys deprived of contact with their mothers appeared to suffer mental distress. The babies became very attached to cheesecloth diapers in their cages, clinging onto them like security blankets. As the story goes, Harlow was on a champagne flight over Detroit in 1957, wondering which would be more powerful, contact comfort or food. Suddenly he got the idea of providing the babies with a choice between two mothers, one of which would provide something soft to embrace, the other of which would provide milk. To make a cuddly mother, Harlow's assistants covered a wire frame with terrycloth. A second mother was identical to the first but had no terrycloth; it was just a wire frame with a built-in milk bottle and nipple. Harlow found that baby monkeys preferred the terrycloth mother, spending as little time as possible on the wire frame mother. Contact comfort was more important than food, except when the baby monkey was actually eating.

Harlow found baby monkeys were very forgiving of their terrycloth mothers. One "mother" was designed to occasionally catapult the baby off, throwing it to the other side of the cage, but the babies always came back. Another had spikes beneath her terrycloth, but the babies put up with the pain. Only running cold water through tubes contained inside the terrycloth mother, making the body cold, made baby monkeys reject it and retreat to a corner of the cage.

PLAY

Because play expends energy and creates a risk of injury, it must have counterbalancing *benefits*, or it would not be positively selected in evolution. What is the adaptive function of play? One hypothesis born out by field studies is that social bonds formed in childhood can be helpful later in life. Smuts, who took the picture of Pandora and Virgil, found friendship did appear to help baboons. Friendships formed in childhood often lasted into adulthood.

Another obvious benefit of play is that it exercises and refines skills to be used in adulthood. Most play is vigorous, repeated, and often spectacular exercise. Repeated performance of motor tasks in mammals results in a physiological "training response" in which muscles and bones hypertrophy (develop), endurance increases, and the economy and precision of the movements increase.

Play is often repetitious, and often the play behaviour is not carried to its usual conclusions (for example, in play-biting the teeth are not closed), or else the play is exaggerated in form (e.g. the bounding of a kitten). *Social* play (pouncing, rough-housing) is often preceded by a stereotyped pose or behaviour that indicates a lack of serious aggression or intention to harm. Rhesus monkeys have a *play-face* they make before bothering an adult, while dogs and cats often crouch halfway before pouncing in play.

Hinde pointed out that play behaviour is often elicited by an older animal. The lioness twitches her tail and cubs play with it. Female chimpanzees tickle and roll their infants while pretending to bite them. Playing with young offspring is evidently pleasurable to adults, too, in moderate amounts, so these behaviours are mutually reinforcing.

REPRODUCTIVE ACTIVITY

Reproductive activity has always been a focus of scientists studying animals because it forms such a prominent part of the behavioural repertoire of all species. Reproduction must occur, if genes are to be passed on. Most acts of most animals are geared, ultimately, toward production of successfully reproducing young. Tinbergen's work with the stickleback, already discussed in this chapter, is only one example of ethological work focusing on the reproductive cycle.

Odor is a widely used form of communication in the animal world, particularly in combination with reproductive activity. Odor bearing molecules are both *durable* and *highly specific*. They convey directional information if carried in the air or water. Therefore they are ideal for marking territory or attracting a potential mate.

Salmon use odor as a homing signal before spawning (laying eggs). Scientists have known for years that salmon return to the stream or pond where they were born during spawning. This drive is so strong that salmon will beat themselves to death against a dam if one is built across a stream between the spawning pond and the sea.

Scientists wondered how salmon found their way back hundreds of miles to the same exact pond where they were born. Did the fish have some sort of mental map that enabled them to follow geographical features? Did they respond to magnetic fields of the earth? The angle of the sun? As it turned out, *odor* was the key. Scientists put a few drops of a chemical with a powerful odor in a spawning pond. When the baby salmon were grown up and returned from the sea, the scientists moved to a new, different pond and put the same chemical there. The salmon went back to this new, different pond. A similar experiment showed that frogs tend to return to a pond that has the same odor as the one where they were born.

Odors called *pheromones* play a role in mating activity. Technically, a pheromone is an *intraspecific distance hormone*, which means it is a chemical used *within a species* (intraspecific) to communicate at a distance. A pheromone is not necessarily a sexual attractant, although that is a common function of pheromones. Prince listed nine responses of insects to odor, including swarming, grooming, exchanging solid foods and exchanging liquids. A few parts per billion of a pheromone floating in the air can trigger the response. For example, male dogs can travel many city blocks to find a female dog that is sexually receptive ("in heat") and emitting pheromones.

Do humans respond to pheromones for sexual attraction? The makers of the fragrance Andron claimed it had pheromone-like effects on the opposite sex, although the chemist who designed Andron said he did not believe it. The vomeronasal organ in the nose of humans may possibly be involved in pheromone-like sexual attraction but so far nobody has proven that a specific substance can trigger an affectionate response in humans.

Social status often has a direct impact on reproductive success. Less dominant animals may be excluded from mating entirely. "Tournaments" are common in many species. The winners get the healthiest, highest ranking mates. That does not prevent lower status animals from trying to reproduce, but they may not be successful. In animals as in humans, it is common for lower status males to "aim too high" and be rejected by desirable females.

SOCIOBIOLOGY

In 1975, a Harvard biologist who specialised in the study of ants—Edward O. Wilson—proposed a new discipline to be called *sociobiology*. It would focus on the biological and evolutionary underpinnings of social behaviour. Wilson wrote a book titled *Sociobiology* in which he proposed genetic explanations of many human social behaviours.

Wilson's book, and his proposed field of study, was immediately attacked by social scientists, including psychologists. Critics like anthropologist Sherry L. Washburn argued that Wilson's ideas were oversimplified and dangerous.

Postulating genes to account for behaviours is a major feature of the application of sociobiology to the interpretation of human behaviours. For example, in the last chapter of *Sociobiology*, genes are postulated to account for more than 25 behavioural situations. There are conformer genes, genes for flexibility, genes predisposing to cultural differences...

... The logic [used by sociologists] is that there must be altruistic genes to account for altruistic acts-just as we learned many years ago that if there are criminal acts there must be criminal genes.

Washburn was alluding to the dark days of eugenics, a pre-WW II theory that was embraced by the Nazis and used to justify the extermination of "undesirable elements" such as criminals, Jews, homosexuals, and mentally retarded people...all on the assumption that they were genetically inferior.

Eugenics first proposed in the 1880s by Sir Francis Galton and was a straightforward but simple-minded extrapolation of Darwinian "survival of the fittest" ideas to the human race. Lack of fitness was equated with criminality, deviance from social norms, and race mixing. For a time, eugenics was endorsed at the highest levels of the British and U.S. governments, by people like Winston Churchill in Britain and President Calvin Coolidge in the United States. Washburn's reference to "criminal genes" in her critique of sociobiology was a reference to eugenics. The Nazi horrors illustrated where such a logic could lead. Related issues still arouse passion in the modern world. In 2007, the Surgeon General of the United States recommended that all pregnant women receive a simple new test for Down Syndrome. Parents of children with Down Syndrome, as well as adults with Down Syndrome, immediately protested that this amounted to a proposal for genocide, because many women would elect to terminate a pregnancy if a fetus tested postitive for Down Syndrome.

Over the last two decades there has been a steady stream of news about researchers discovering "the gene" that links people to diabetes, Alzheimer's, obesity, schizophrenia, depression and many other afflictions. Yet most of those hard-wired gene-disease links— as many as 95 percent of them, according to one British study published in 2003— don't hold up to closer scrutiny. Instead, follow-up studies find that if there is any measurable genetic link to these common diseases, it results from the more complex interactions of many genes with one another, as well as with the environment.

If you examine that statement carefully, it does not say that 95% of the gene-disease links are false; it says that things always turn out to be more complex than implied by the headlines. True research findings of subtle genetic influences lead to falsely simplistic news reports implying that a single stretch of DNA is causing something to happen. Genetic influences are seldom that direct.

However, evolution does have an impact on the brain and behaviour, including social behaviour. The evolutionary perspective on social behaviour is now taken for granted by most psychologists, and the field of evolutionary psychology is thriving. Only the label "sociobiology" itself became unpopular. In 1997 the journal *Ethology and Sociobiology* changed its name to *Evolution and Human Behaviour*. Edward O. Wilson accused them of intellectual cowardice.

Wilson himself remained influential among psychologists. In 1999 he was the keynote speaker at the American Psychological Association convention a few months before New. Years Day 2000. The convention addressed future trends, and it is no coincidence Wilson played a prominent role. He discussed his book *Consilience: The Unity of Knowledge*, which urged the integration of all natural sciences, including biology, ethology, and psychology.

EVOLUTIONARY INFLUENCES ON HUMAN BEHAVIOUR

Students of animal behavior have been applying concepts from ethology to human behavior for decades. Tinbergen and Lorenz saw ways in which their concepts applied to humans, and their ideas were popularized by authors like Desmond Morris, who wrote a bestseller in 1967 called The Naked Ape, full of speculations about how various species-typical behaviors of humans might have evolved.

Lorenz's student Iraneous Eibl-Eibesfeldt made human ethology his specialty. He applied the field observational techniques of ethology to humans, recording behavior of people in native cultures with the side-viewing technique described in Chapter 1. Eibl-Eibesfeldt wrote a textbook called Human Ethology that introduced ethological concepts and showed how they applied to humans.

The 1990s saw the publication of many books and articles using the phrase evolutionary psychology. Like sociobiology and human ethology, this discipline interprets human behavior as a manifestation of underlying neural circuitry modified by evolution. Evolutionary psychologists explain common tendencies of humans by analyzing their adaptive value. Topics studied by evoluitionary psychologists include the following:

Factors Influencing Mate Selection

Researchers have found that people generally find statistically average faces more attractive than faces with extreme features. Humans (and animals) prefer symmetry of the body and face in a potential mate. Not surprisingly, people also like healthy skin and consider it more attractive than blemished or diseased skin.

Nonverbal and Automatic Forms of Social Communication

Humans, like other animals, have species-typical displays. Eibl-Eibesfeldt studied the eyebrow flash (briefly raised eyebrows) that serves as a recognition signal and invitation to interaction (but don't do it in Japan, where it is considered crude and sexual). Eibl-Eibesfelt verified an observation made by Margaret Mead, the anthropologist: that young girls of many cultures use a stereotyped flirtation display (a glance, followed by looking away, covering the face, giggling, and looking back). Eckhard Hess showed that pupil size telegraphed the emotional state of a human and altering the pupil size in photographs would change people's evaluation of pictures.

Paul Ekman became famous for documenting basic, universally interpretable human facial expressions found in all cultures. By studying the anatomy of facial muscles, he went on to catalogue the full range of possible facial expressions, of which only a subset are frequently used by any culture. David R. Buss and his laboratory in Texas has a long record of studying unconscious social signalling among young dating-age adults, such as the "neck cant" (a sexy tilt of the head, common in females), fiddling with hair (common when a person wants to impress somebody of the opposite sex) and other signals of body language and posture. Many of Buss's observations were made unobtrusively by his research assistants at places like bars and beaches where young single people meet and assess each other.

Simulations and Models of Evolutionary Processes

Some psychologists use mathematical or computer models to predict how individuals with particular traits could interact in large populations. The impact of traits on the population as a whole can be modeled on a computer, given certain starting assumptions. Game theory can be used to study social cooperation or competition, negotation or predation.

Adaptive Function

What is the adaptive function of a yawn? For primates, it is a dominance display. However, it also cools the brain. Competing explanations sometimes lead to competing predictions that can be tested with research, or it could be (as in this example) that a behavior turns out to have multiple functions.

Evolutionary Explanations

Honor killings and jealous rages by spurned males may have "worked" in the ancestral environment. Women who stay with an abuser may be acting out an ancient script that was once genetically advantageous. The Stockholm Syndrome (in which kidnapped people bond with their captors) may have been an adaptive response in ancient times. Somnomania may have protected males from sneak attacks at night.

Evidence for Evolutionary Influences on Human Behaviour

Speculations about evolution and behavior are, in a sense, too easy. Anybody can conjure up explanations for why a currently-existing pattern is adaptive. It is all post hoc or after-the-fact (the behavior exists, therefore it must be adaptive), and nobody can disprove it. Are there better ways to test the claims of evolutionary psychologists?

Authors from the early to mid 20th Century, such as Carl Jung, Tinbergen, Lorenz, Eibl-Eibesfelt and popularizers like Morris, relied on universality as their main argument. If a pattern (such as the eyebrow flash, or attraction to made-up faces) appeared in culture after culture, in widely separated places on earth, then it might come from the human biological blueprint. That is still a compelling argument in the case of truly ubiquitous (universal) human behavioral patterns. Presumably they are ubiquitous for a reason.

Other standards of proof now exist and are more likely to result in journal publications. For example:

1. *Straightforward evidence of genetic control.* Now that DNA can be mapped and altered on a location-by-location basis, behavioral genetics (the field that relates behavior to genes) is becoming more precise and definitive. If knocking out or introducing

a gene produces a reliable consequence in behavior, then this is clearly a point at which evolutionary change could be made through mutation or recombination. Another avenue of change is through environmental influences that modify gene expression while leaving the underlying DNA untouched, an effect called epigenetics. Epigenetic factors include dietary influences on obesity, activity level, and health. Epigenetic factors may be passed on to children, so this is a whole new level of complexity to consider for modern scientists studying genetic influences.

2. *Systematic observational evidence.* In his classic study of the Yanomami Indians of the Amazon, Chagnon reported that warriors (despite the risk of injury and death) reproduced at a higher rate than men who chose not to fight. Warriors were preferred as mates, married more often, and produced more children. This finding is controversial, but it is an example of an evolutionary argument relevant to behavior and backed by replicable evidence that could be studied directly or in computer simulations.

   Eibl-Eibesfelt's video archives are another important form of observational evidence. His recordings show behavioral patterns of daily village life, in a variety of native cultures all around the world. This is the best way to bolster arguments about universality or cultural variations of human behavior: simply collect data from a variety of relatively unspoiled native cultures (such as they exist...Eibl-Eibesfeldt went about his task with urgency because such cultures were disappearing rapidly).

3. *Replicable experiments.* Classic findings about non-verbal social communication lend themselves to laboratory study by psychologists, often in college settings using subject pools drawn from modern student populations. For example, students can be asked to rate faces for attractiveness. This type of research can be used to document widespread human preferences for such things as facial symmetry, "average" features, large pupil size, and clear healthy skin, although researchers must always take care not to overgeneralize from student populations.

4. *Simulations.* The effects of different strategies of competition or cooperation can be modeled on computers or simulated in laboratory games pitting one participant against another. The effect of genetic costs or benefits on the population numbers of a species can be simulated on a computer and the effects of varying assumptions about the ecosystem can be tested.

Cautions about Evolutionary Thinking

With all the solid research supporting evolutionary influences on behavior, it is perhaps no wonder that evolutionary perspectives are achieving widespread acceptance within the discipline of psychology. However, students of evolutionary psychology should keep in mind several cautionary points:

What are a few complications in relating genes to behavior?

— Almost all behavior is the result of multiple genetic influences (i.e. most behavior is polygenic).

— A single genetic change almost always results in multiple behavioral changes. This is called pleiotropy. It is important because all the changes produced by a genetic alteration (not just the most obvious) can affect evolution.

— A behavior that evolved in the past to serve one function may, in later times, serve entirely different functions.

The last point is sometimes called evolutionary opportunism. Evolution always works with what is already there. A behavior or body part that is already existing may have previously unexploited uses or benefits. If so, it can fall under a new set of selective pressures.

For example, ground-dwelling birds may use a pre-flight movement in their mating displays, even though members of the species can no longer fly. Some time in their ancestral past, the pre-existing behavioral pattern (pre-flight movement of the wings) was assimilated into the mating display. Why? Because it was available in the male bird's behavioral repertoire, and some females liked it. Now it serves a new function: attracting a mate. Consequently, the genes that control muscle development and expression of that behavior are maintained by new selective pressures. In other words, the behavior serves a new purpose (helping to attract a mate) and that is what keeps a flight movement programmed into the species DNA of a flightless bird. Such a behavior is said to be emancipated from its original function.

The twin concepts of opportunism and emancipation greatly complicate the interpretation of evolved behavior. One cannot assume the present function of a behavior (or body alteration) is the same as its ancestral function.

For example, dreaming may have evolved to serve some basic biological purpose unrelated to cognition, but in humans the vivid nighttime experiences may have influenced daytime behavior. If the influences were positive (aiding differential reproduction) then dreaming may have taken on a new function for humans, such as helping with problem solving. In that case, the occurrence of meaningful dreaming could be an example of opportunism: emergence of a new function for an existing behavior.

Dreaming is probably not emancipated from its original function, because it probably still serves its original purpose as well as occasionally offering guidance for decision-making. But the newer function of solving problems (and giving warnings, etc.) may have emerged well after the original function of dreaming in pre-human mammals. Warnings and inspirations from dreams can have considerable impact on reproductive success, so dreaming may now be adaptive for humans in several different ways: aiding memory, guding decisions, keeping people out of trouble during night hours, and many other possible functions.

The above factors are all complications of evolutionary thinking. Keeping them in mind can help a student avoid simplistic thinking. In addition, students should avoid several illogical ways of discussing evolution.

Evolution should not be reified or treated as a thing or a force of nature. Evolution is just a word used to label all the various factors that influence differential reproduction and propagation of DNA. Evolution is not a thing or an agent with a separate existence. Therefore it is erroneous to speak of evolution making something happen. That type of language is a sloppy shorthand for saying that a genetic change leads some creatures to prosper and reproduce while others fail to pass on their genes. Evolutionary change may be "blind" (unpredictable) but it is not random in a statistical sense. Consider how babies resemble their biological parents. The outcome is not predictable, but neither is it random.

In statistics, a random process in one in which all outcomes are equally likely. However, when a biological system is altered (even by a random process like a cosmic ray) the results are not random in a statistical sense. The evolutionary past of any system makes some outcomes much more likely than others. The range of possible outcomes (when genetic expression is altered) always depends on which components and arrangements already exist in the system. Those pre-existing structures have been cumulating and modifying for millions of years. (In the same sense, a human never has a truly "random" thought...all thoughts are influenced by a person's history of learning, etc.)

At the same time as the details of an evolutionary process are unpredictable, the general shape or outcome of an evolutionary process can be predicted, and when a given mutation or behavior is sufficiently adaptive, it may emerge independently several different times. Therefore one cannot assume that similar structures or behaviors are produced by the same ancestral process. Sometimes identical behaviors in different population groours are analogous (resulting from similar evolutionary processes) rather than homologous (resulting from a common ancestors). One cannot distinguish analogous vs. homologous processes on the basis of observation alone. DNA analyses are required to resolve the issue of common ancestry.

Convergent evolution is the term used to label strongly constrained, repeated evolutionary outcomes. Some adaptations appear more than once, as the result of independent processes that differ in their details. An adaptation will appear very reliably if there is a sufficiently compelling network of constraints or selective pressures.

When humans starting domesticating cattle, for example, it became highly adaptive for adult humans to tolerate lactose, so adults could benefit from drinking milk from their cattle. This was a life or death matter during famines and droughts. Lactose tolerance in adulthood required only a simple genetic modification to preserve an enzyme already produced in babyhood. Consequently, as shown by DNA evidence, lactose tolerance evolved five different times in the past 2-3,000 years, independently, in human populations of northern Africa. That is an example of convergent evolution.

Ironically, the parallel between learning and evolution has been discovered repeatedly by different scientists. Pringle, Ashby, Donald T. Campbell, Jean Piaget, and B.F. Skinner all noticed it. Both learning and evolution proceed by variation and selective retention of adaptive variants, the evolutionary pattern. It is certainly fitting that this insight occurred independently to so many scientists, because it emphasizes the parallel between learning and evolution itself. Convergent evolution can occur in scientific discoveries (a form of learning) just as it can in biological evolution.

Human Ancestral Environment

In both humans and other animal species, evolved behavior patterns reflect the selective pressures of the ancestral environments. Evolutionary psychologists, following John Bowlby, sometimes speak of the EEA or environment of evolutionary adaptedness. This is the environment in which an evolved tendency was adaptive. In other words, it is the environment in which a behavior evolved. A simpler term is ancestral environment, if the term is understood to refer to the period of time when a behavior evolved, not earlier or later times.

For humans, the EEA was the environment of the most recent 125,000 years or so. Lactose tolerance, the example used above, evolved only recently, in the past 2,000 years. That is extremely recent by evolutionary standards, and it shows the great advantage milk-drinking provided to people in northern climates. Many human behaviors presumably evolved during a much earlier period of pre-history.

DNA studies show that biologically modern humans emerged from a small group of approximately 1,000 individuals who lived about 100,000 years ago. The population bottleneck (ours was an endangered species at the time) made the human race genetically uniform. Geneticists say there is less variation between the DNA of humans from opposite ends of the earth than there is between gorillas or chimps from adjacent patches of jungle. The reason is that gorillas and chimpanzees experienced no population bottleneck in their recent evolutionary past, so their populations have had more time to accumulate random changes and variations. Humans, by contrast, are all very similar in their DNA.

Organized agriculture began relatively recently, about 10,000 to 20,000 years ago. After that, many groups of modern humans lived in villages. However, cultural change occurs much faster than biological change. Most human behavioral characteristics were already in place by the time humans learned to domesticate plants and animals. Most evolutionary psychologists believe the dominant social environment for evolving humans was the older, hunting-and-gathering, nomadic way of life.

Impact of Environmental Change

The adaptive value of a behavior, or any other trait, is always relative to some environment. When the environment changes, so does the adaptive value of a particular behavior.

On the famous Galapagos Islands where Darwin made many of his observations, the Woodpecker Finch uses twigs and cactus spines to probe for insects in holes of dead trees. This is an excellent example of an adaptive behavior in a particular environment. It represents an adjustment to the environment of the islands, where insects in the trees provide food for the finches, which gives the finches an evolutionary advantage. By exploiting this food source, they are more likely to survive and reproduce than finches that cannot use twigs in this way.

However, the finch's behavior is adaptive only because there are no woodpeckers on the island. If there were woodpeckers in the area, the finch's behavior would no longer give it a reproductive advantage. Woodpeckers are more efficient at extracting insects from dead wood, so food would not be available for the twig-using finch, and it might become extinct.

To interpret the fitness or adaptive value of a behavior, one must specify the environment. That includes the other creatures present. The introduction or extinction of a single new species can tip the balance of survival for dozens of other species in that ecosystem. For exapmle, cats introduced to island ecosystems by humans have a devastating effect. They are excellent predators and the animals living on isolated islands have no evolved defenses against them. Fire ants have devastated competing ant species in the southern U.S. In Europe, the arrival of modern humans around 40,000 years ago probably caused the extinction of the Neanderthals, who had lived there for 300,000 years. Again and again, the fossil record shows that the arrival of modern humans in an ecosystem is followed by the mass extinction of large animals in the same area, probably due to hunting. Early humans are now believed to have hunted the wooly mammoth to extinction.

Humans are often so successful at modifying their environments that they bring about the destruction of their own societies. The Maya civilization burned wood to cook limestone, which was ground up and used to make a form of concrete, the main construction material for their massive ceremonial temples. Simple calculations show that this consumption of forest resources was unsustainable. Eventually all the trees were gone, and that (along with a drought and a belief system predicting the end of the world) led to the end of Mayan civilization.

A similar process occurred on Easter Island, where an isolated population of humans cut down the native species of trees until they were all gone. The ecosystem was devastated and the humans eventually faced repeated famines. Many similar events are occurring today, for example, in fishing villages of the northeastern United States, where a traditional way of life is dying because the level of fishing has reduced stocks to the point where people can no longer make a living by fishing.

On issue after issue—including some of the most significant problems of modern times—the challenge facing our species is to modify our behavior intelligently so we can encourage cultural and behavioral practices that are healthy and sustainable, while discontinuing those that are self-defeating. Evolutionary psychology may help. In some cases, simulations may suggest adaptive strategies for the future. In many other cases, evolutionary psychology can help by explaining the origins of destructive behavior patterns left over from the past. If we understand where they came from and how they might have been adaptive in ancient times, it may help us to move beyond them and not to continue blindly with practices that are maladaptive in the modern world.

Tinbergen and Lorenz emphasized solitary animals or pairs of animals, in their classic work of the 1930s through 1950s. With the 1960s and 1970s came a new emphasis on group interactions of animals, such as competition, communication, and predatory/prey relationships. Predator/prey competition led to specialized adaptations in many animals. Bats are such efficient predators of insects that many insects have adaptations specially designed to help them avoid bats. Other prey animals evolved colors that warned potential predators of a bad taste, or imitated stimuli feared by the predator animal.

A variety of examples show how visual patterns are used by insects to ward off potential predators. Many moths have eyespots on their wings, because birds (which prey on moths) are themselves preyed upon by animals with large staring eyes (like primates, cats, and owls). Birds are highly sensitive to visual patterns but relatively insensitive to the size of a pattern, so visual mimicry in insects works by scaring away predator birds.

Alarm responses are often shared between different species, benefiting all. In the 1980s, researchers discovered that vervet monkeys had at least three different alarm calls for different types of predators-a simple form of communication. Since then, the same phenomenon of multiple, specific alarm calls has been discovered in many species. Humans also participate in shared alarm calls (with watchdogs), and humans themselves have a stereotyped alarm call: the scream. Struggles for territory and social status are common in group living animals. Monkeys, rams, and porpoises are known for aggressive combat to establish who is dominant. Animals use a variety of signals to avoid damaging combat when possible. Threat displays often involve what Tinbergen called "intention movements." Submission postures are used to indicate deference to higher status animals, in group living species. Appeasement displays are used to calm down a threatening animal.

Prosocial behaviors are friendly interactions of group-living animals. Grooming is one such behavior, often elicited with invitation displays. For example, monkeys turn their backs to other monkeys to elicit grooming. Harry Harlow discovered that baby rhesus monkeys separated from their mothers would cling to a surrogate (substitute) mother made of soft cloth. He said there was a need for "contact comfort."

Most animals have highly developed action patterns devoted to courtship, mating, nest building, and raising the young. Odors are often involved in identifying a home territory, for example, when salmon return to the stream where they were born, for spawning.

Social ethology has generated many concepts that can be applied to humans. The discipline of sociobiology, proposed by Edward O. Wilson in 1975, was very controversial because simplistic references to genetic influences. It reminded people of eugenics, a discredited movement of the late 1800s and early 1900s. However, evolutionary psychology slowly gained credibility, and by the 1990s and 2000s it was a vigorous sub-discipline of psychology. There is ample evidence for the influence of evolution on human behavior, and the ideas of evolutionary psychology can be tested in replicable forms of research.

Evolutionary reasoning applied to human being is tricky for several reasons. Genetic and epigenetic influences are complex and subtle, often involving multiple interactions. Evolved behaviors, once established in a population, may be used opportunistically in new ways that meet different needs. For humans, cultural changes are very rapid, and behaviors that may have been adaptive in ancient times may not be sustainable in the modern world.

REFERENCES

Alcock, J. 2001. *Animal Behavior: An Evolutionary Approach* (7 ed.) Sinauer Associates Inc., Massachusetts.

Menzer, J. 1991. *Animal Learning in the 21st Century*. Buffalo, NY: Bose-Jarman.

Hailman JP. 1977. *Optical Signals: Animal Communication and Light*. Indiana Univ Press, Bloomington.

Scott, J.P. and Fuller, J.L. 1965. *Genetics and the Social Behavior of the Dog*. University of Chicago Press.

Wilson EO. 2000, *Sociobiology: The New Synthesis*, Twenty-fifth Anniversary Edition, Belknap Press.

9

Human Behaviour Patterns

Human behaviour is the collection of behaviours exhibited by human beings and influenced by culture, attitudes, emotions, values, ethics, authority, rapport, hypnosis, persuasion, coercion and/or genetics. The behaviour of people (and other organisms or even mechanisms) falls within a range with some behaviour being common, some unusual, some acceptable, and some outside acceptable limits.

In sociology, behaviour is considered as having no meaning, being not directed at other people and thus is the most basic human action. Behaviour should not be mistaken with social behaviour, which is more advanced action, as social behaviour is behaviour specifically directed at other people. The acceptability of behaviour is evaluated relative to social norms and regulated by various means of social control.

The behaviour of people is studied by the academic disciplines of psychology, sociology, economics, and anthropology. In 1970, a book was published called "The Social Contract: A Personal Inquiry into the Evolutionary Sources of Order and Disorder" written by the anthropologist Robert Ardrey. The book and study investigated animal behaviour and then compared human behaviour as a similar phenomenon.

Humans are biological creatures, as much as crocodiles, cougars, and capabara. We are the product of millions of years of evolution, our physical make-up changing to make us fitter to survive and reproduce. However, although humans are animals, we also have something that no other animal has: the most complex social structure on Earth. We gather in families, tribes, clans, nations. We have an incredibly sophisticated method of interacting—speech. We can communicate over time and distance through printing and broadcasting. Our memories are the longest, our interactions the most intricate, our perception of the world simultaneously the broadest and most detailed.

The combination of biology and society is what makes us what we are and do what we do. Biology guides our responses to stimuli, based on thousands of generations of ancestors surviving because of their responses. Our social structures dictate restrictions on and alterations in how we carry out our biological responses.

Neither biology nor society stands without the other. For some people, this is a contradiction—either nature (biology) controls people, or nurture (society) does. But in fact we filter everything through both to determine how we react to stimuli. The following is a discussion of the two sides of human nature: first, the biological basis of our responses to the world around us, and second, the social factors that affect those responses and make us human.

BIOLOGICAL BASIS OF HUMAN BEHAVIOUR

The three main elements biology contributes to human behaviour are:

1) Self-preservation;
2) The reason for self-preservation, reproduction; and
3) A method to enhance self-preservation and reproduction, greed.

Let us discuss each in turn.

Self-preservation

Self-preservation is keeping yourself alive, either physically or psychologically. The latter includes mentally or economically healthy.

A lioness slowly, stealthily, works through the tall grass toward the herd of wildebeest. A doe, unaware of the danger lurking in the grass, separates slightly from the herd. With a rush, the lioness bursts into a run to take down the doe. The startled doe bounds away, running and swerving, trying to escape. The lioness, unable to keep up the pace, gives up, and the doe escapes back into the herd.

To be successful as a species, the members of that species must have a desire to survive long enough to pass on their genes to offspring. A species with a death-wish dies out rather quickly. Those species that don't die out have members that have devoted some attention to staying alive long enough to have young. It is from those individuals and therefore species that all living things are descended.

The desire to stay alive is an instinctive one, built into the psyche of the organism. The organism will seek those elements of its environment that will enhance its chances for survival. These include food, water, oxygen, and periods of rest to allow the body to repair any wear and tear on the tissues.

Alternately, it will avoid or evade those elements that might reduce its chances for survival. Such dangers include predators, starvation, dehydration, asphyxiation, and situations that can cause damage to the body.

These seek or avoid drives influence the behaviour of organisms: iron seeking bacteria will move toward magnetism, gnus will migrate hundreds of miles to find new pastures, a human will resort to cannibalism; an amoeba will flow away from an electric current, an antelope will run from a lion, a human will obey a killer or withstand torture.

Survival Through Evolution

A phrase that has often been misquoted, "Survival of the Fittest," actually means survival of the fit. The term fit mean an organism has those attributes that allow it to get the most out of its environment: gather food, drink, oxygen, rest, sex. The better it is at doing this, the more fit it is.

A niche is a position within an environment that calls for certain attributes to exploit that environment. An environment can contain any of a variety of elements: amount of water, from ocean to desert; type of land, from marsh mud to solid rock; amount of vegetation, from none (the Arctic and Antarctic) to abundant (rainforests). It can also contain animal life, from the tiniest insects to blue whales and everything in between. It is the combination and degree of each of these elements that create niches.

As an example, let's look at just one of these elements. Say there are many small animals, like mice, in an area. A small carnivore like a wildcat could find a lot of food. Thus, it would fit into this niche and thrive. However, when the number of mice decreases, the wildcat can find less food, and has a lesser chance of survival.

If the wildcat has competition from other small carnivores, like foxes, the one that is particularly good as a predator, through cunning or speed or some other attribute, will catch more food. This lessens the amount of food available for the competition, and thus drives the competition out. If the fox is better at catching mice than the wildcat, the wildcat will either die or have to move to another niche in which it will be the better predator. On the other hand, if there are no small animals but many big animals, like antelope, neither a fox nor a wildcat would have much success preying on them. Thus, they wouldn't fit in such a niche. However, large carnivores such as lions would.

Of course, nothing stays the same forever. Niches alter through geologic, climatic and, in the present day, man-made changes in land, water and air. A volcano can create a new island. An ice age can lock up hugh quantities of water in ice caps and glaciers, creating areas of land where oceans once rolled. Continental drift can push seabeds to the tops of mountains. Humans can chop down forests and build cities. All these changes alter the niches, the environmental conditions under which the life in those niches live.

Of course, this means the life has to change as well, to match the new conditions. If it doesn't, it dies. An example is a moth in England. It was originally a mottled white, which allowed it to blend into the light bark of the trees in its area. However, in the 19th century factories in this area began to belch out soot from their chimneys that settled on the trees, changing the tree bark from mottled white to mottled black. The moth could no longer blend in and thus was easy prey to birds. However, some of the moths were darker and thus less noticeable. After a few generations of these darker moths surviving and passing on their genes, the standard color changed to mottled black, and the moth, now blending into the dark bark, survives.

Note that such changes are not conscious decisions made by the organism: the moth did not say to itself, "The bark is getting dark—I'd better change color, too." It is simply that there are variations between individuals in any species (an advantage of sexual reproduction and its combining of genes). Some of those variations are detrimental: the dark moth variations were easy prey when the tree bark was light. However, as the conditions in a niche change, those same variations can become advantageous, enhancing rather than weakening chances for survival.

Such changes in an organism's physical characteristics are, of course, accidental. If no variations exist in a species that contribute to survival when conditions change, or if conditions change too quickly for advantageous variations to be passed on to enough descendants, the species can die out.

Survival Through Strategy

Other changes in an organism can develop over time. These are survival strategies, rather than physical changes, that improve the organism's chances for survival. For example, some animals have perfected the technique of hibernating during periods when the food supply is low. Marmots have developed a social structure that provides lookouts who watch for predators and sound a warning when one appears. Prairie dogs dig their burrows with multiple entrances and exits so if a predator comes in one door, the dogs can leave through another.

These survival strategies are adaptations to niche conditions, but unlike physical changes are not necessarily genetic changes. Such strategies as hibernation, of course, require genes that alter the animal's physiology to slow heartbeat, lower body temperature, and otherwise decrease its metabolism. Others are instinctive, hardwired genetically into the animal's brain, such as a fawn's curling up and freezing when predators are about.

However, some survival strategies are learned behaviours. That is, the young learn them from older animals that learned them from their ancestors. For example, most predators teach their young the techniques of successful hunting. In general, it appears the higher the complexity of the nervous system of the animal, the more likely strategies are learned rather than instinctive. Sharks, with a relatively simple nervous system, hunt by instinct and need no instruction on how to go about it. Lions, with a complex system, must learn the techniques of stealth, stalk, and attack.

Again, in most animals, the strategies are not conscious decisions, but responses to stimuli such as hunger, thirst, asphyxiation, fear, or exhaustion. If conditions change so the instinctive strategy is dangerous rather than beneficial, the animal can die. For example, the fawn's freeze response to fear would be deadly if there was no cover to hide in while frozen. The musk ox strategy is to form a stationary circle with the young in the centre and the older members facing outward, rather than running away. This is excellent against wolves, but deadly when faced with spears and guns (perfect, however, for the human survival strategy of group hunting with weapons). The musk ox cannot consciously decide that this strategy isn't working and that they must try another.

The combination of genetic and learned responses to stimuli create an animal's reaction to stimuli. For example, the genetically dictated instinctive reaction to a threat to self-preservation is the "fight or flight" syndrome. When threatened, an animal undergoes several physiological changes that have become genetically hardwired into the animal's body. The changes include an increased rate of respiration to provide more oxygen to the muscles, an accelerated heart beat to speed up the blood flow, a lessening in sensitivity to pain, and changes in the blood stream, including an injection of adrenalin and diversion away from the organs to the muscles. These physiological changes prepare the animal to either fight for survival or run away from danger.

However, learned responses can mitigate the instinctive, depending on the complexity of the animal's nervous system. That complexity increases an animal's options in reacting to stimuli. For example, an amoeba will avoid an electric field automatically—an instinctive reaction unmitigated by a survival strategy. A starving rat, however, will run across an electrified grid that gives it painful shocks if there is food on the other side. It can learn a survival strategy—the shocks, though causing the instinctive fight-or-flight physiological changes, aren't going to kill it. Starvation will.

Self-preservation and Humans

All the above applies to humans as much as any other animal: humans desire personal survival; seek food, drink, rest, sex; fit into niches; must adapt to changing conditions.

Humans are subject to the same stimuli and reactions as any other animal. Hunger, thirst, asphyxiation, fear, and exhaustion are physical sensations that cause instinctive physical reactions. Most of these reactions are unpleasant, and people avoid the stimuli that cause them, or, if they're unavoidable, take actions to reduce them. Thus you eat when hungry, drink when thirsty, fight for air, run from dangerous situations, sleep. In any case, the reactions are good in that they tell you you're in a situation that could result in injury or death. These responses are instinctive, and we have no more control over them than we do over our eye color.

Actually, we do have control over our eye color. The reason we do is why our approach to self-preservation is different from all other creatures. We have a brain that is capable of perceiving and solving problems. We change our eye color with contact lenses. We react to a threatening situation through applying our brains to the problem and finding a solution to it.

The difference between humans and other animals is that, unlike any other animal (as far as we know), we can and do consciously respond or alter our response to a stimulus. The greatest example lies in the existence of amusement parks, where people deliberately subject themselves to stimuli that any other creature on earth would go to great lengths to avoid. Imagine, if you can, the reaction of a dog to a roller coaster. If it didn't leap out at the first movement, it would cringe in bottom of the car until it probably had a heart attack. Yet, humans go on such rides for fun, our minds accepting that the ride is safe, and thus control the terror such a thing would cause in any other creature.

Indeed, the physical manifestations of the stress of the workplace, such as ulcers, headaches, nervous breakdowns, is often considered a result of the fight or flight syndrome at work on the body, while the mind is required to remain under stimuli that no other creature would willing accept. For example, being bawled out by your boss would, in another animal, cause a fight or the chastisd to run. Humans, though, stand, listen, nod their heads, say "yes, I understand" and go back to work (probably muttering uncomplimentary comments about the boss under their breath).

Even more, humans can alter rather than merely adapt to the environments in which we find ourselves to enhance our chances for survival. The invention of agriculture and the domestication of animals improved the food supply; the building of dwellings enhanced shelter from the elements; science and medicine have greatly increased human lifespan and the quality of that life. Human ingenuity has altered every aspect of the world to enhance the human life.

However, humans live in an extremely complex society. Thus, self-preservation is a much more complicated proposition than among other animals. Eating to satisfy hunger is more than just finding proper vegetation or hunting; shelter for rest and recuperation is more than finding a convenient cave or nest; avoiding predators is difficult because it is often hard if not impossible to tell what is a predator (the only real predators on humans are other humans). Even avoiding dangerous situations (such as car crashes) is difficult because of human technology. Things can happen so quickly danger isn't apparent until it's too late to do anything about it.

To deal with the complexity, human society has become, to a large extent, an economic one. That is, the connections between unrelated people is often based on distribution of resources.

Reproduction

Nature has designed life, from the smallest microbe to the largest whale, to have one basic purpose. That purpose is to reproduce. Whatever else an organism does, reproduction is its basic goal. All else, from dolphins frolicking to the Pyramids, is gravy.

An organism is a combination of elements such as carbon, iron, hydrogen, potassium, sodium, etc.. How the elements combine into molecules, and how those molecules are arranged, determines what the organism is, whether a bacterium, bedbug or buffalo. The instructions for this arrangement are in the organism's genes. They provide the guidelines for everything from eye color (or even if there are eyes and if so how many) to the hard-wired instincts on how to react to stimuli. To make a copy of itself, the organism must create another organism that contains the same genes.

Certainly most living things go out of their way to stay that way. They devote a lot of time and attention to self-preservation. However, what basic reason is there for staying alive? If nature just wanted life, everything could be immortal—just do it once and leave it at that. But nothing is immortal; everything eventually wears out (or more accurately for life, dies).

For life to continue, it must make fresh copies of itself. The purpose of self-preservation is to stay alive long enough to make those fresh copies. If an individual's death contributes to that individual's reproduction, it dies. For example, the male preying mantis is eaten by the female while inseminating her, providing her with nourishment she needs to produce and lay her eggs. However, in his sacrifice he passes on that part of his life that's important to nature—his genes.

Almost everything an organism does enhances its ability to stay alive, at least long enough to reproduce. An extreme example is the Pacific salmon, which devotes years to staying alive. It then engages in "kamikaze reproduction," (technically, semelparity) literally committing suicide in its efforts to reproduce. All of its activities, from eating to dodging predators to travelling thousands of miles, aim at that final, fatal return to its original stream to spawn.

Other things that an organism does are attempts to increase its chances at reproducing. For example, among animals that reproduce sexually, males compete with each other to gain access to females. The competitions may be physical, such as butting heads, or psychological, by having the loudest roar. In any case, the competition is to get the status or prove the fitness that will entice the female to choose the winner.

The more complex an organism, the more of its environment it can apprehend and try to manipulate to its own advantage. Often it does this through cooperation with other organisms. Humans, the ultimate in environmental manipulators, do more things, with more others, to more effect, than any other creatures on earth. You may watch television to relax; relaxation can benefit your health; the healthier you are the longer you may live; the longer you live the greater your chance to reproduce.

Do humans think this way when they sit down to watch Gilligan's Island? Of course not. But the body might be saying, "Take a strain off," and, with many ways to relax, television may be the choice. Painting pictures may be relaxing, or provide money to buy food and shelter or prove fitness to a potential mate. Looking at the pictures may be soothing, or provide a more relaxing atmosphere, or be an investment. Jogging can improve health, and perhaps provide opportunities to meet healthy potential mates.

The earth teems with thousands of species of creatures, all doing their best to continue teeming. As discussed above, a species continues by reproducing, by making copies of the members in that species. Species that don't reproduce, for whatever reason, don't survive.

Samuel Butler once said, "The hen is the egg's way of making another egg." There is a much truth in this adage. In the natural world, the egg's vehicle, be it fish, fowl or flower, doesn't matter. What does matter is that the vehicle gets the egg fertilised, hatched, and far enough along in development to have a chance of producing another egg. The egg contains the hen's genes, which combine with the rooster's genes. The genes create from the raw materials of organic life another chicken. That chicken then has another egg. It is not the egg, but the genetic material that the egg nourishes and brings to fruition that is important. It is the genes that must be passed on if a species is not to become extinct.

In addition, the egg wants to have another egg as much as possible like itself. A species is a collection of genetically similar individuals. Since it's the genetic material, the genes, that makes each egg an individual member of a species, the egg wants to pass its own genes on to the next generation.

Of course, to nature the individual isn't important; what is important is the species to which the individual belongs. An examination of nature shows this to be true. Herbivores that are susceptible to predation often herd in large numbers. Fish and insects may eggs in the thousands and millions. Plants produce seeds and spores in the millions and billions. In this way, although individual members of a species may fall to predators or bad luck, the species is safe from extinction.

However, this does not mean that the individual doesn't care about reproduction. In fact, it is the reason for the individual's existence. It isn't the species that carries the genes, it's the individuals in that species that do. This desire to pass on its own genes is why an individual is so concerned with self-preservation. It must personally survive to pass on its genes.

An organism wants to pass its own genes on to offspring. If it didn't want to, it wouldn't care if it reproduced or not. Enough of these individuals, and the species dies out. However, that event is unlikely. Without an instinct to pass on its own genes that individual doesn't, and its lack of an instinct to reproduce dies with it. Those individuals that have the instinct are the ones that continue the species, and therefore breed the instinct into future generations.

Asexual Reproduction

Some may think that reproduction is synonymous with sex. After all, most of the species we see (or pay any attention to) reproduce that way. However, asexual reproduction is also very common. Asexual reproduction is the replication of an organism. The organism can fission, bud, or clone itself. Fissioning is the organism splitting itself in two parts, each part containing the same genes. For example, an amoeba fissions, dividing itself into two individuals, each containing the same genes. The hydra, a sea animal that looks like a plant, buds off pieces of itself that grow into new hydras, all with the same genes. The honeybee can reproduce through parthenogenesis, developing new bees containing only the queen's genetic material.

There are advantages to asexual reproduction. First, genes rarely mutate. Thus, a genetically successful (that is, adapted to its niche) organism doesn't develop mutations that may lessen its success. Second, it devotes all its reproductive energy to creating offspring. No time or energy has to be devoted to the finding and courting a mate. There are, of course, disadvantages to asexual reproduction. If there weren't, there would be no point in sex. However, obviously there is a point to sex—it mitigates asexuality's disadvantages.

Sexual Reproduction

Much of terrestrial reproduction is through bipolar sexual activity, the mating of male and female. This way each sex passes on its share of genes to any resulting offspring. In sexual reproduction, each sex contributes one half of the genes that go into creating a new organism. The male contributes his half in sperm, basically a small packet of genes with a tail that lets it swim to the female's contribution. She contributes an egg that contains her genes and a quantity of food for the developing offspring. The sperm and egg combine through sexual intercourse.

The advantages of sexual over asexual reproduction are many. First, of course, is that a sexual species can evolve faster than an asexual one. A sexual species combines genes randomly. Thus, each offspring is a unique individual, a combination of two other individuals. This individuality may include a new variation of some aspect. This is a mutation. If niche conditions change, this new variation (or some other) may be more successful in exploiting the new conditions. Interestingly, sexual species can also resist evolving. A mutation in an asexual species passes on to any descendant; there isn't any choice. But in a sexual species a mutation can be "edited" out, since half of the maternal or paternal genes aren't passed on. If the mutation isn't passed on, it disappears. However, there is a more substantial difference between asexual and sexual reproduction than what each gender contributes, whether it's a sperm or an egg, or its value in evolutionary terms. There is the "cost of sex."

Reproductive Strategies

Nature designs reproductive strategies to produce the strongest, smartest, most viable organisms. They in turn reproduce and continue the chain of life. Continuing the chicken and egg analogy, to reproduce the genes in the female's egg combine with the genes in the male's sperm. This is done through sexual intercourse. Since this is so, individual females and males of each specie must want sexual intercourse with each other. Without such desire, nature's reproductive strategy of passing genes from one generation to another might fail. Thus, sexual desire, very simplistically, is built in to every organism. In addition, the egg wants (in a metaphorical sense) to combine the best possible genes to improve the next chicken's (and therefore the next egg's) chances at survival and reproduction. Animals' reproductive strategies revolve around these two needs—mating, and mating with the best possible partner.

Sexual desire is an instinctive reaction in animals. If the desire isn't there, neither is the possibility of reproduction. If it is there, it is based on an individual's perception of a suitable mate. That perception usually is a set of criteria that the opposite sex must meet, or at least approximate. An individual regards a member of the opposite gender that meets or exceeds (at least, within the available pool of possibilities) those criteria as being superior to others of that gender. It is therefore the individual that provides the chance for the highest quality offspring with the best chance of survival. When the criteria are not met, sexual desire does not occur.

The criteria can include coming into heat, showing strength, clearly being healthy, providing a safe environment for offspring, being readily available as a sexual partner, some other factor or factors, or any combination. It all depends on what the organism's reproductive strategy deems important. Please always bear in mind that, in all but humans (as far as we know) there is no conscious element to reproductive strategy—instinctive reaction guides it. Such criteria must, of course, be reciprocal. Simply because she meets his criteria does not mean he meets her's, and vice versa. It is when both partners' criteria overlap enough for both partners to accept them that mating takes place.

Again, the chicken and the egg: the hen's criteria for sexual desire are those that would allow the best combination of genes. If the rooster is good enough to satisfy the criteria, it is superior to one that does not meet them. Thus, its qualities are those that should combine with her's and pass on, and he's the rooster with which she mates. Because of the costs of sex, the criteria a potential mate must meet or approximate differ greatly, depending on whether the animal applying them is male or female.

Males

Most males are promiscuous. Genetically, it is the most practical course of action. The more females with which a male mates, the greater number of offspring containing his genes are possible. In addition, the cost of sex in terms of time and energy is considerably lower for the male than the female. It is therefore in the male's (and thus the male's genes') best interest for the male to mate with as many females as he can. Thus, a male's criteria that a female should satisfy can be quite simple:

1) the female must be healthy to carry the fetus to term and support it after birth until it reaches self-sufficiency or puberty;

2) she must be young enough for greater viability, health, and ability to support offspring long enough after birth for it to reach self-sufficiency or puberty;

3) impregnable condition.

As is apparent, all of the above criteria are physical and usually obvious to the male's eyes or nose. In other words, a male can quickly see and/or smell whether a female fulfils his criteria. For example, a female cockroach releases pheramones (a chemical scent) when she's sexually ready that draws males from everywhere. The female chimpanzee's sexual organs swell and redden when she's receptive, and the males line up in hopes of mating with her. Nothing else is important—to the male.

In addition, the sex act is of paramount importance to the male; it is how he impregnates the female. She must therefore be receptive to him. After that, his concern is having more females to impregnate. He may try to ensure that his already mated females don't mate with any other males, but the more successful he is with the former, the less he probably worries about (if those are the words for something he doesn't think about at all) the latter.

Females

In most species, females bear the brunt of the cost of sex in both time and energy: up to millions of calories and years of time. Among mammals, she must not only produce the young, she must rear them to the point of self-sufficiency. Thus, unlike the male, she doesn't have the choice of promiscuity, of creating as many offspring as possible as quickly as possible; she cannot abandon offspring as soon as they are born, or her genes die with the infant.

All this means she must be highly selective in her choice of mates if she wishes to produce the highest quality offspring in her reproductive lifetime. If she selects just any male that comes along, she could waste all the time and energy that pregnancy and rearing require on a possibly weak or nonviable offspring. Her criteria thus are aimed at getting the best possible male. What is important is the quality of genes he brings and the help, if any, she will have while carrying, bearing and rearing her young.

Her criteria, therefore, are more complex than the male's. Not only must he be physically acceptable, but should satisfy other factors that may contribute to her and her offspring's welfare. These can include leadership, status within a group, and fighting skill. It could even include the male's ability to put up with handicaps, such as the peacock's conspicuous and undoubtedly unwieldy tail. The sex act, and his participation, being so brief, doesn't have to be of any great interest to her. He need merely be able to achieve orgasm.

These two seemingly contradictory attitudes toward reproduction, promiscuity in the male and selectivity in the female, must be reconciled. If they aren't, the species will die out due to lack of offspring. Species, therefore, devise reproductive strategies that bring males and females together.

Strategies are many and varied. They can include singing, displays of physical characteristics, dancing, pheromone release, flashing lights, or combat. In most species, it is the male that does something to attract the female's attention. There is a reason for that.

The first aspect of virtually any strategy is that the female makes the choice when it comes to mating. She has such a high biological stake in the outcome of mating that she must select the best possible male. If he does the choosing, a species could waste its reproductive strength, which resides in the female and the time and energy she puts in, on nonviable young. Since it is sound reproductive strategy for her to be the arbiter, he can only apply for permission to mate and wait for her decision.

How the male applies for permission to mate is a major part of any strategy. The application is to attract her attention and convince her that he's the male she should choose because he fulfils her criteria, whatever they might be. For example, male lightning bugs flash into the night, hoping a female will respond. The peacock opens one of the most beautiful displays in nature, a monster fan of iridescent tail feathers. A male bower bird constructs an elaborate well-decorated mating bower of grass, then dances while she looks on. She also checks out the bowers of other males. She mates with the one she thinks best, while the losing males try to attract another female.

Among mammals, the same strategy applies. He applies, she decides. His application is often by challenging other males to mating rights: elk bellow; rams butt heads; moose wrestle using their huge palmate antlers; elephant seals fight, sometimes to the death.

These psychological or physical battles are to win her approval by winning the combat. For example, the African lion bases its reproductive strategy on combat to take over a pride and thus gain a harem. Male lions fight each other to prove their strength, strength to pass on to offspring. The strongest lion wins the right to reproduce. When a new lion takes over a pride, he kills cubs sired by the loser, ensuring two results. First, the pride wastes no strength and resources on raising inferior cubs. Second, the lioness goes into estrus and becomes impregnable again. Female lions will mate only with the winners of the combat. They deem the winner as the one most likely to sire strong offspring that have the best chance of survival. However, if she still doesn't approve, she can walk away, or give him a quick (and painful) swat to let him know she's not interested.

In the wolf pack, only the alpha male and female reproduce. The alphas are the leaders of their pack. They have shown through combat and leadership that they are the strongest and smartest of the wolves in the pack. The rest of the pack work to raise the alphas' pups. When either of the alphas fails, the male through being defeated in a fight, the female in producing poor pups or getting too old, new alphas replace them and the pack raises the new alphas' pups.

Not only predatory animals use the combat strategy for determining the best male to sire young. Among most herbivores, such as deer, elk, gnus, sheep, horses and cattle, the males battle to gain the right to reproduce (this is, of course, in the wild). Once a male proves his dominance, the females will mate with him, as many females as he can keep with him and away from other males.

The female strategy is to wait, watch, and choose. It is rare for the female to pursue the male. Exceptions appear in those species in which the male is the nurturing parent rather than the female. In such species, like the phalarope and the seahorse, the male devotes the time and energy to the rearing of young. He thus becomes a "resource" for the female in propagating her genes, as, in most species, the female is a resource for the male. Thus, the female fights off other females until she lays her eggs, then goes to find another mate.

Greed

Greed has an extremely negative connotation for most people. It conjures up images of Ebenezer Scrooge and Shylock, chortling over their gold and ignoring the plights and miseries of others. However, it is actually the gathering of resources, the more the better. Biologically, for any organism that is successful greed is good. Any form of life must gather resources that allow it to survive and reproduce. The resources may be food, water, sunlight, minerals, vitamins, shelter. Without these things, the organism dies. Since the two most basic purposes of life are to live and to reproduce, it should do everything it can to avoid dying through a lack of resources.

Greed is one organism getting a larger piece of the pie, more of the necessary resources, than other organisms. For example, in the Amazonian rain forest, an occasional tree dies and falls. This leaves an opening to the sun in the continuous canopy of foliage. Plants and trees race each other to grow into that opening. The winners in the race fill the hole; the losers die through lack of sunlight. The greed for sunlight means life.

Again, as for self-preservation and sex, greed is an instinctive reaction. When presented with resources, the instinct is to grab them, use them, take advantage of them. This isn't a conscious decision. An animal, when starving, wants more food; when thirsty, more water. If it means taking it from another animal, that's what it does if it can.

You may ask, what about those animals who feed their offspring, though they're starving themselves? Remember that the second purpose of life is to reproduce. This requires not only producing the young. Once it's born it must be kept alive until it's self-sufficient. If it dies, then all the time, effort and energy to produce it must be repeated to produce another one. However, once it reaches self-sufficiency the parent's genes will, most likely, be passed on to another generation. Keeping the offspring alive, even at the expense of the parent dying, is of paramount importance. Thus, a parent caring for its young at its own expense is not an act of selflessness; it's an act of genetic selfishness.

SOCIAL BASIS OF HUMAN BEHAVIOUR

We are the only creatures on earth (as far as we know) that can remember the past as discrete events, then connect those events with present conditions. Then, on the basis of those connections, we can consciously decide what to do, and project possible present actions into the future consequences of those actions. Thus, unlike other animals that react to stimuli as they occur, humans live not only in the present, but in the past and the future. A dog may bristle at a threat, but not at a threat that's long gone or hasn't occurred yet. Humans will do all three. It is this ability to remember the past, relate it to the present, and project into the future that is a special province of humans. This ability allows us to manipulate our environment, communicate across distance and time, and evolve incredibly complex societies and cultures.

The ability to think also means that we have two ways of viewing things, instinctively and intellectually. The instinctive view is instant and uncontrollable: it's the way the mind is wired, unalterable except by evolution. The intellectual view is learned and alterable, based on the culture and society in which the person lives. We can use it to mitigate the effects of the instinctive view (although when instinct comes in conflict with intellect, the conflict can be so great that the result is neurosis or even psychosis).

Self-preservation: Sociological Basis

Staying alive is a personal quest for any animal. It is personal survival that allows it to continue its genetic line. However, an animal doesn't necessarily have to survive on its own. Another aspect of personal survival is the forming of social groups within a species. When staying alive is not just the responsibility of the individual, but other members of the species help the individual to survive, and vice versa, all members' chances are enhanced.

Social groups come in all levels, from couples to herds, from two to thousands. The purpose of a social group and the level it takes is often dictated by how well it serves to promote the survival of the members. For example, dik dik antelope form pairs, while gnus form herds of thousands. The dik dik lives in heavily forested areas, filled with thickets and under brush. This provides plenty of hiding places for individuals, but not groups. Thus, the fewer animals there are, the less likely a predator will find one.

Gnus, on the other hand, live on the wide-open African savannahs. In these conditions, there is virtually no place for an individual to hide. However, the individual can hide behind and within a large group, and the larger the group the more individuals that can be protected from predators. Thus, more animals in the group mean fewer are vulnerable to predation. There are, of course, groups that fall in all levels between these two extremes. Monkeys form bands of thirty or so, depending on their habitat. Generally, the more open the area, the more members in the band, forming a compromise between hiding in the habitat and hiding in the group.

Again, it appears the higher the neural complexity of the animal, the more the social group is used to enhance survival. In addition, the higher the neural complexity of the animal, the higher the complexity of the society. The more complex the society, the more that society devotes its resources to an individual's welfare. For example, alligators will help their own young to survive. They won't help, and may even eat, other alligators and their young. Most herbivores, on the other hand, form herds, where members, related or not, can support each other against predators, or at least provide hiding places. Lions form prides and wolves form packs, helping each other in hunting and raising young. Humans will adopt children, and help complete strangers stay alive.

Humans have the most complex society of any creature on earth, which means we extend self-preservation beyond personal physical survival. We live in extremely complex and interdependent societies, where people band together in groups for mutual aid and protection. Such groups include families, friendships, associations, tribes, clans, states, nations. The members of these groups work together to help each other. Also, since the group enhances the members' chances of survival, group survival means personal survival. The individual benefits by supporting the group, because the group reciprocates by supporting the individual.

This is clear for most animals. Wolf packs and lion prides hunt together, allowing them to get more, and bigger, game. Marmots and prairie dogs post lookouts to see danger and warn the other members of the group. Chimpanzees organise hunting bands.

Such cooperation strategies increase the resources necessary to survival that the group can gather. More members can kill more and larger game. More members can take and defend a larger territory, increasing food resources and defensive positions against predators. This helps satisfy the group's immediate survival needs.

Of course, resources don't last forever. Food rots. Weather or fire can destroy shelter such as trees and thickets. Water can dry up or shift course. When this happens, the group hunts again, moves, or dies.

Humans have gone through the same process. During our evolution from primate to person we formed bands, hunting groups, mutual protection societies. Each member contributed to and shared in resources.

However, humans have developed societies that go beyond the needs of the present. The human ability to think makes us consider future needs as well, and how the individual and the group can survive. For example, because we can remember what we've experienced in the past, we realise that food rots. When it does, and if we can't find food in the future, we'll starve. We realise that shelter can burn down, that water can dry up or get polluted. To take care of future needs, we realise there must be a way of transporting present resources into that future in a way that isn't perishable.

Human ingenuity has found such a way of transporting present resources such as food, drink, shelter, mutual protection, etc. into the future. Any effort that a member of a group does that helps that group survive is recompensed in a way that the work is acknowledged later. These resources are converted into a symbolic representation of them: money. For example, farmers grow food and sell it for money. When they in turn need food or any other resource, they reconvert the money into what they need . People use money for all aspects of physical survival: buying food, shelter, clothing, medical services, protection. Whatever the individual or the group needs to survive is converted into and out of its symbolic representation.

Of course, the human ability to think, which provides so many advantages for survival, carries with it a disadvantage. Being able to think about the future means being able to worry about it as well. Most people imagine what their future will be like. They project current conditions into the future, usually based on past experience. It is normal for people to be concerned whether they will have the resources they need. However, worry is when such extrapolations concentrate on negative results. This can have a negative effect on a person's psychological health.

Money is once again the major resource about which people worry—it's how people can get the other resources they need. The fear of not having enough money for current needs, or thinking there won't be enough for future needs, can create the same physiological effects as any other threat. It can cause the stress of the fight-or-flight syndrome. Sufficient stress, particularly unrelieved stress, can be as dangerous to personal survival as a predator.

However, just as money can relieve the stress of personal survival, it can relieve psychological stress. We can buy insurance that will replace lost resources and relieve the worry about not having enough in the future; we can buy entertainment to relieve mental stress; we can buy counseling when the stress becomes overwhelming. Again, whatever the individual or the group needs to survive is converted into and out of its symbolic representation.

Humans, unlike other animals, also have a conscious awareness of the effects of reproduction, that our progeny is our posterity. We therefore extend self-preservation to our children to an even greater degree than other animals. We care for them not just until self-sufficiency or puberty, but well beyond. This increases their chances of survival, and thus of our genes' survival in the future.

Finally, our ability to remember the past and project into the future means we may help total strangers, those that have no genetic relationship to us at all. We can remember debts for days, months, and years. We can also imagine we or our children might need help in the future. We extend this sense of obligation and return far beyond that of any other creature. We don't expect that a return will necessarily come from the specific person we helped. We just believe that helping will result in help when needed. Thus, by helping others, any others, we help ourselves.

BEHAVIOURAL MODERNITY

Behavioural modernity is a term used in anthropology, archeology and sociology to refer to a list of traits that distinguish present day humans and their recent ancestors from both living primates and other extinct hominid lineages. It is the point at which Homo sapiens began to demonstrate a reliance on abstract thought and to express cultural creativity. These developments are often thought to be associated with the origin of language.

There are two main theories regarding when modern human behaviour emerged. One theory holds that behavioural modernity occurred as a sudden event some 50 kya, possibly as a result of a major genetic mutation or as a result of a biological reorganization of the brain. Proponents of this theory refer to this event as the Great Leap Forward or the Upper Paleolithic Revolution.

The second theory holds that there was never any single technological or cognitive revolution. Proponents of this view argue that modern human behaviour is basically the result of the gradual accumulation of knowledge, skills and culture occurring over hundreds of thousands of years of human evolution. Proponents of this view include Stephen Oppenheimer in his book Out of Eden, and John Skoyles and Dorion Sagan in their book Up from Dragons: The evolution of human intelligence.

Modern human behaviour is observed in Cultural universals which are the key elements shared by all groups of people throughout the history of man. Examples of elements that may be considered cultural universals are language, religion, art, music, myth, cooking, games, and jokes. While some of these traits distinguish Homo sapiens from other species in their degree of articulation in language based culture, they all have analogues in animal ethology. Since cultural universals are found in all cultures including some of the most isolated indigenous groups, scientists believe that these traits must have evolved or have been invented in Africa prior to the exodus.

Classic evidence of behavioural modernity includes:

— finely made tools,

— fishing,

- evidence of long-distance exchange or barter among groups,
- systematic use of pigment (such as ochre) and jewellery for decoration or self-ornamentation,
- figurative art (cave paintings, petroglyphs, figurines)
- game playing and music
- burial

A more terse definition of the evidence is the behavioral B's: blades, beads, burials, bone toolmaking, and beautiful. It might be thought that behavioral modernity preceded language, but the complex behaviors from the list above are thought to suggest language was necessary and that they must have been at least contemporary developments.

Timing

Whether modern behavior emerged as a single event or gradually is the subject of vigorous debate.

Great leap forward

Advocates of this theory argue that the great leap forward occurred sometime 50-40kya in Africa or Europe. They argue that humans who lived before 50kya were behaviorally primitive and indistinguishable from other extinct hominids such as the Neanderthals or Homo erectus.

Proponents of this view base their evidence on the abundance of complex artifacts, such as artwork and bone tools of the Upper Paleolithic, that appear in the fossil record after 50kya. They argue that such artifacts are absent from the fossil record from before 50kya, indicating that earlier hominids lacked the cognitive skills required to produce such artifacts.

Jared Diamond states that humans of the Acheulean and Mousterian cultures lived in an apparent stasis, experiencing little cultural change. This was followed by a sudden flowering of fine toolmaking, sophisticated weaponry, sculpture, cave painting, body ornaments, and long-distance trade. Humans also expanded into hitherto uninhabited environments, such as Australia and Northern Eurasia.

The Great Leap Forward was concurrent with the extinction of the Neanderthals, and it has been suggested that Cro-Magnon interaction with Neanderthals caused this extinction.According to this model, the emergence of anatomically modern humans predates the emergence of behaviorally modern humans by over 100kya.

Continuity hypothesis

Proponents of continuity hypothesis hold that no single genetic or biological change is responsible for the appearance of modern behavior. They contend that modern human behavior is the result of sociocultural and sociobiological evolution occurring over hundreds of thousands of years. Continuity theorists base their assertions on evidence of modern behavior that can be seen in the Middle Stone Age (approximately 250-50kya) at a number of sites in Africa and the Levant. For example, a ritual burial with grave goods at Qafzeh is Middle Stone Age(MSA) having been dated to 90kya. The use of pigment is noted at several MSA sites in Africa dating back more than 100kya.

Continuity theorists believe that what appears to be a technological revolution at the onset of the Upper Paleolithic is most likely a result of increased cultural exchange resulting from a growing human population. Some continuity theorists also argue that the rapid pace of cultural evolution during the Upper Paleolithic transition may have been triggered by adverse environmental conditions such as aridity arising from glacial maxima. They further dispute that anatomical modernity predates behavioral modernity, stating that changes in human anatomy and behavioral changes occurred stepwise.The findings of Curtis Marean and his colleagues of fishing and symbolic behavior dating to 164,000 years ago on the southern African coast strongly support this analysis.

EMOTIONS

An emotion is a mental and physiological state associated with a wide variety of feelings, thoughts, and behavior. Emotions are subjective experiences, or experienced from an individual point of view. Emotion is often associated with mood, temperament, personality, and disposition. The English word 'emotion' is derived from the French word émouvoir. This is based on the Latin emovere, where e- (variant of ex-) means 'out' and movere means 'move'. The related term "motivation" is also derived from movere.

No definitive taxonomy of emotions exists, though numerous taxonomies have been proposed. Some categorizations include:

— 'Cognitive' versus 'non-cognitive' emotions
— Instinctual emotions (from the amygdala), versus cognitive emotions (from the prefrontal cortex).
— Basic versus complex: where base emotions lead to more complex ones.
— Categorization based on duration: Some emotions occur over a period of seconds (e.g. surprise) where others can last years (e.g. love).

A related distinction is between the emotion and the results of the emotion, principally behaviours and emotional expressions. People often behave in certain ways as a direct result of their emotional state, such as crying, fighting or fleeing. Yet again, if one can have the emotion without the corresponding behaviour then we may consider the behaviour not to be essential to the emotion. The James-Lange theory posits that emotional experience is largely due to the experience of bodily changes. The functionalist approach to emotions (e.g. Nico Frijda) holds that emotions have evolved for a particular function, such as to keep the subject safe.

Classification

Basic and complex categories, where some are modified in some way to form complex emotions (e.g. Paul Ekman). In one model, the complex emotions could arise from cultural conditioning or association combined with the basic emotions. Alternatively, analogous to the way primary colors combine, primary emotions could blend to form the full spectrum of human emotional experience. For example interpersonal anger and disgust could blend to form contempt.

Robert Plutchik proposed a three-dimensional "circumplex model" which describes the relations among emotions. This model is similar to a color wheel. The vertical dimension represents intensity, and the circle represents degrees of similarity among the emotions. He posited eight primary emotion dimensions arranged as four pairs of opposites. Some have also argued for the existence of meta-emotions which are emotions about emotions., "Meta-emotions".

Another important means of distinguishing emotions concerns their occurrence in time. Some emotions occur over a period of seconds (e.g. surprise) where others can last years (e.g. love). The latter could be regarded as a long term tendency to have an emotion regarding a certain object rather than an emotion proper (though this is disputed). A distinction is then made between emotion episodes and emotional dispositions. Dispositions are also comparable to character traits, where someone may be said to be generally disposed to experience certain emotions, though about different objects. For example an irritable person is generally disposed to feel irritation more easily or quickly than others do. Finally, some theorists (e.g. Klaus Scherer, 2005) place emotions within a more general category of 'affective states' where affective states can also include emotion-related phenomena such as pleasure and pain, motivational states (e.g. hunger or curiosity), moods, dispositions and traits.

Theories

Theories about emotions stretch back at least as far as the Ancient Greek Stoics, as well as Plato and Aristotle. We also see sophisticated theories in the works of philosophers such as René Descartes, Baruch Spinoza and David Hume. Later theories of emotions tend to be informed by advances in empirical research. Often theories are not mutually exclusive and many researchers incorporate multiple perspectives in their work.

Somatic theories

Somatic theories of emotion claim that bodily responses rather than judgements are essential to emotions. The first modern version of such theories comes from William James in the 1880s. The theory lost favour in the 20th Century, but has regained popularity more recently due largely to theorists such as António Damásio, Joseph E. LeDoux and Robert Zajonc who are able to appeal to neurological evidence.

James-Lange theory

William James,in the article 'What is an Emotion?', argued that emotional experience is largely due to the experience of bodily changes, the Danish psychologist Carl Lange also proposed a similar theory at around the same time, so this position is known as the James-Lange theory. This theory and its derivatives state that a changed situation leads to a changed bodily state. As James says 'the perception of bodily changes as they occur IS the emotion.' James further claims that 'we feel sorry because we cry, angry because we strike, afraid because we tremble, and neither we cry, strike, nor tremble because we are sorry, angry, or fearful, as the case may be.'

This theory is supported by experiments in which by manipulating the bodily state, a desired emotion is induced. Such experiments also have therapeutic implications (e.g. in laughter therapy, dance therapy). The James-Lange theory is often misunderstood because it seems counter-intuitive. Most people believe that emotions give rise to emotion-specific actions: i.e. "I'm crying because I'm sad," or "I ran away because I was scared." The James-Lange theory, conversely, asserts that first we react to a situation (running away and crying happen before the emotion), and then we interpret our actions into an emotional response. In this way, emotions serve to explain and organize our own actions to us.

Neurobiological theories

Based on discoveries made through neural mapping of the limbic system, the neurobiological explanation of human emotion is that emotion is a pleasant or unpleasant mental state organized in the limbic system of the mammalian brain. If distinguished from reactive responses of reptiles, emotions would then be mammalian elaborations of general vertebrate arousal patterns, in which neurochemicals (e.g., dopamine, noradrenaline, and serotonin) step-up or step-down the brain's activity level, as visible in body movements, gestures, and postures. In mammals, primates, and human beings, feelings are displayed as emotion cues.

For example, the human emotion of love is proposed to have evolved from paleocircuits of the mammalian brain (specifically, modules of the cingulated gyrus) designed for the care, feeding, and grooming of offspring. Paleocircuits are neural platforms for bodily expression configured millions of years before the advent of cortical circuits for speech. They consist of pre-configured pathways or networks of nerve cells in the forebrain, brain stem and spinal cord. They evolved prior to the earliest mammalian ancestors, as far back as the jawless fishes, to control motor function.

Presumably, before the mammalian brain, life in the non-verbal world was automatic, preconscious, and predictable. The motor centers of reptiles react to sensory cues of vision, sound, touch, chemical, gravity, and motion with pre-set body movements and programmed postures. With the arrival of night-active mammals, circa 180 million years ago, smell replaced vision as the dominant sense, and a different way of responding arose from the olfactory sense, which is proposed to have developed into mammalian emotion and emotional memory. In the Jurassic Period, the mammalian brain invested heavily in olfaction to succeed at night as reptiles slept — one explanation for why olfactory lobes in mammalian brains are proportionally larger than in the reptiles. These odor pathways gradually formed the neural blueprint for what was later to become our limbic brain.

Emotions are thought to be related to activity in brain areas that direct our attention, motivate our behavior, and determine the significance of what is going on around us. Pioneering work by Broca (1878), Papez (1937), and MacLean (1952) suggested that emotion is related to a group of structures in the center of the brain called the limbic system, which includes the hypothalamus, cingulate cortex, hippocampi, and other structures. More recent research has shown that some of these limbic structures are not as directly related to emotion as others are, while some non-limbic structures have been found to be of greater emotional relevance.

Cognitive theories

There are some theories on emotions arguing that cognitive activity in the form of judgements, evaluations, or thoughts is necessary in order for an emotion to occur. This, argued by Richard Lazarus, is necessary to capture the fact that emotions are about something or have intentionality. Such cognitive activity may be conscious or unconscious and may or may not take the form of conceptual processing. An influential theory here is that of Lazarus. A prominent philosophical exponent is Robert C. Solomon). The theory proposed by Nico Frijda where appraisal leads to action tendencies is another example. It has also been suggested that emotions (affect heuristics, feelings and gut-feeling reactions) are often used as shortcuts to process information and influence behaviour.

Perceptual theory

A recent hybrid of the somatic and cognitive theories of emotion is the perceptual theory. This theory is neo-Jamesian in arguing that bodily responses are central to emotions, yet it emphasises the meaningfulness of emotions or the idea that emotions are about something, as is recognised by cognitive theories. The novel claim of this theory is that conceptually based cognition is unnecessary for such meaning. Rather the bodily changes themselves perceive the meaningful content of the emotion because of being causally triggered by certain situations. In this respect, emotions are held to be analogous to faculties such as vision or touch, which provide information about the relation between the subject and the world in various ways. A sophisticated defense of this view is found in philosopher Jesse Prinz's book Gut Reactions and psychologist James Laird's book Feelings.

Affective Events Theory

This a communication-based theory developed by Howard M. Weiss and Russell Cropanzano (1996), that looks at the causes, structures, and consequences of emotional experience (especially in work contexts.) This theory suggests that emotions are influenced and caused by events which in turn influence attitudes and behaviors. This theoretical frame also emphasizes time in that human beings experience what they call emotion episodes - a "series of emotional states extended over time and organized around an underlying theme" This theory has been utilized by numerous researchers to better understand emotion from a communicative lens, and was reviewed further by Howard M. Weiss and Daniel J. Beal in their article, Reflections on Affective Events Theory published in Research on Emotion in Organizations in 2005.

Cannon-Bard theory

In the Cannon-Bard theory, Walter Bradford Cannon argued against the dominance of the James-Lange theory regarding the physiological aspects of emotions in the second edition of Bodily Changes in Pain, Hunger, Fear and Rage. Where James argued that emotional behaviour often precedes or defines the emotion, Cannon and Bard argued that the emotion arises first and then stimulates typical behaviour.

Two-factor theory

Another cognitive theory is the Singer-Schachter theory. This is based on experiments purportedly showing that subjects can have different emotional reactions despite being placed into the same physiological state with an injection of adrenaline. Subjects were observed to express either anger or amusement depending on whether another person in the situation displayed that emotion. Hence the combination of the appraisal of the situation (cognitive) and the participants' reception of adrenaline or a placebo together determined the response. This experiment has been criticized in Jesse Prinz (2004) Gut Reactions.

Component process model

A recent version of the cognitive theory comes from which regards emotions more broadly as the synchronization of many different bodily and cognitive components. Emotions are identified with the overall process whereby low-level cognitive appraisals, in particular the processing of relevance, trigger bodily reactions, behaviors, feelings, and actions.

Disciplinary approaches

Many different disciplines have produced work on the emotions. Human sciences study the role of emotions in mental processes, disorders, and neural mechanisms. In psychiatry, emotions are examined as part of the discipline's study and treatment of mental disorders in humans. Psychology examines emotions from a scientific perspective by treating them as mental processes and behavior and they explore the underlying physiological and neurological processes. In neuroscience sub-fields such as affective neuroscience, scientists study the neural mechanisms of emotion by combining neuroscience with the psychological study of personality, emotion, and mood. In linguistics, the expression of emotion may change to the meaning of sounds. In education, the role of emotions in relation to learning are examined.

Social sciences often examine emotion for the role that it plays in human culture and social interactions. In sociology, emotions are examined for the role they play in human society, social patterns and interactions, and culture. In anthropology, the study of humanity, scholars use ethnography to undertake contextual analyses and cross-cultural comparisons of a range of human activities; some anthropology studies examine the role of emotions in human activities. In the field of communication sciences, critical organizational scholars have examined the role of emotions in organizations, from the perspectives of managers, employees, and even customers. A focus on emotions in organizations can be credited to Arlie Russell Hochschild's concept of emotional labor. The University of Queensland hosts EmoNet, an email distribution list comprised of a network of academics that facilitates scholarly discussion of all matters relating to the study of emotion in organizational settings. The list was established in January, 1997 and has over 700 members from across the globe.

In economics, the social science that studies the production, distribution, and consumption of goods and services, emotions are analyzed in some sub-fields of microeconomics, in order to assess the role of emotions on purchase decision-making and risk perception. In criminology, a social science approach to the study of crime, scholars often draw on behavioral sciences, sociology, and psychology; emotions are examined in criminology issues such as anomie theory and studies of "toughness", aggressive behavior, and hooliganism. In law, which underpins civil obedience, politics, economics and society, evidence about people's emotions is often raised in tort law claims for compensation and in criminal law prosecutions against alleged lawbreakers (as evidence of the defendant's state of mind during trials, sentencing, and parole hearings). In political science, emotions are examined in a number of sub-fields, such as the analysis of voter decision-making.

In philosophy, emotions are studied in sub-fields such as ethics, the philosophy of art (e.g., sensory-emotional values, and matters of taste and sentiment), and the philosophy of music. In history, scholars examine documents and other sources to interpret and analyze past activities; speculation on the emotional state of the authors of historical documents is one of the tools of interpretation. In literature and film-making, the expression of emotion is the cornerstone of genres such as drama, melodrama, and romance. In communication studies, scholars study the role that emotion plays in the dissemination of ideas and messages. Emotion is also studied in non-human animals in ethology, a branch of zoology which focuses on the scientific study of animal behavior. Ethology is a combination of laboratory and field science, with strong ties to ecology and evolution. Ethologists often study one type of behavior (e.g. aggression) in a number of unrelated animals.

Perspectives on Emotions from Evolution Theory

Perspectives on emotions from evolution theory were initiated in the late 19th century with Charles Darwin's book The Expression of Emotions in Man and Animals. Darwin's original thesis was that emotions evolved via natural selection and therefore have cross-culturally universal counterparts. Furthermore, animals undergo emotions comparable to our own . Evidence of universality in the human case has been provided by Paul Ekman's seminal research on facial expression. Other research in this area focuses on physical displays of emotion including body language of animals and humans (see affect display). The increased potential in neuroimaging has also allowed investigation into evolutionarily ancient parts of the brain. Important neurological advances were made from these perspectives in the 1990s by, for example, Joseph E. LeDoux and António Damásio.

American evolutionary biologist Robert Trivers argues that moral emotions are based on the principal of reciprocal altruism. The notion of group selection is of particular relevance. This theory posits the different emotions have different reciprocal effects. Sympathy prompts a person to offer the first favor, particularly to someone in need for whom the help would go the furthest. Anger protects a person against cheaters who accept a favor without reciprocating, by making him want to punish the ingrate or sever the relationship. Gratitude impels a beneficiary to reward those who helped him in the past. Finally, guilt prompts a cheater who is in danger of being found out, by making them want to repair the relationship by redressing the misdeed. As well, guilty feelings encourage a cheater who has been caught to advertise or promise that he will behave better in the future.

Sociology of Emotions

We try to regulate our emotions to fit in with the norms of the situation, based on many – sometimes conflicting – demands upon us which originate from various entities studied by sociology on a micro level – such as social roles and 'feeling rules' the everyday social interactions and situations are shaped by – and, on a macro level, by social institutions, discourses, ideologies etc. For example, (post-)modern marriage is, on one hand, based on the emotion of love and on the other hand the very emotion is to be worked on and regulated by it. The sociology of emotions also focuses on general attitude changes in a population. Emotional appeals are commonly found in advertising, health campaigns and political messages. Recent examples include no-smoking health campaigns and political campaign advertising emphasizing the fear of terrorism.

Emotions and Psychotherapy

Depending on the particular school's general emphasis either on cognitive component of emotion, physical energy discharging, or on symbolic movement and facial expression components of emotion, different schools of psychotherapy approach human emotions differently. While, for example, the school of Re-evaluation Counseling propose that distressing emotions are to be relieved by "discharging" them - hence crying, laughing, sweating, shaking, and trembling. Other more cognitively oriented schools approach them via their cognitive components, such as rational emotive behavior therapy. Yet other approach emotions via symbolic movement and facial expression components (like in contemporary gestalt therapy).

MOTIVATIONS

Motivation is the set of reasons that determines one to engage in a particular behavior. The term is generally used for human motivation but, theoretically, it can be used to describe the causes for animal behavior as well. This article refers to human motivation.

According to various theories, motivation may be rooted in the basic need to minimize physical pain and maximize pleasure, or it may include specific needs such as eating and resting, or a desired object, hobby, goal, state of being, ideal, or it may be attributed to less-apparent reasons such as altruism, morality, or avoiding mortality.

A reward, tangible or intangible, is presented after the occurrence of an action (i.e. behavior) with the intent to cause the behavior to occur again. This is done by associating positive meaning to the behavior. Studies show that if the person receives the reward immediately, the effect would be greater, and decreases as duration lengthens. Repetitive action-reward combination can cause the action to become habit. Motivation comes from two things: you, and other people. There is extrinsic motivation, which comes from others, and intrinsic motivation, which comes from within you.

Applying proper motivational techniques can be much harder than it seems. Steven Kerr notes that when creating a reward system, it can be easy to reward A, while hoping for B, and in the process, reap harmful effects that can jeopardize your goals.

Rewards can also be organized as extrinsic or intrinsic. Extrinsic rewards are external to the person; for example, praise or money. Intrinsic rewards are internal to the person; for example, satisfaction or a feeling of accomplishment.

Some authors distinguish between two forms of intrinsic motivation: one based on enjoyment, the other on obligation. In this context, obligation refers to motivation based on what an individual thinks ought to be done. For instance, a feeling of responsibility for a mission may lead to helping others beyond what is easily observable, rewarded, or fun.

A reinforcer is different from reward, in that reinforcement is intended to create a measured increase in the rate of a desirable behavior following the addition of something to the environment.

Intrinsic and Extrinsic Motivation

Intrinsic motivation

Intrinsic motivation occurs when people engage in an activity, such as a hobby, without obvious external incentives. This form of motivation has been studied by social and educational psychologists since the early 1970s. Research has found that it is usually associated with high educational achievement and enjoyment by students. Intrinsic motivation has been explained by Fritz Heider's attribution theory, Bandura's work on self-efficacy, and Ryan and Deci's cognitive evaluation theory. Students are likely to be intrinsically motivated if they:

— attribute their educational results to internal factors that they can control (e.g. the amount of effort they put in),

— believe they can be effective agents in reaching desired goals (i.e. the results are not determined by luck),

— are interested in mastering a topic, rather than just rote-learning to achieve good grades.

In knowledge-sharing communities and organizations, people often cite altruistic reasons for their participation, including contributing to a common good, a moral obligation to the group, mentorship or 'giving back'. In work environments, money may provide a more powerful extrinsic factor than the intrinsic motivation provided by an enjoyable workplace.

In terms of sports, intrinsic motivation is the motivation that comes from inside the performer. That is, the athlete competes for the love of the sport.

Extrinsic motivation

Extrinsic motivation comes from outside of the performer. Money is the most obvious example, but coercion and threat of punishment are also common extrinsic motivations.

In sports, the crowd may cheer the performer on, and this motivates him or her to do well. Trophies are also extrinsic incentives. Competition is often extrinsic because it encourages the performer to win and beat others, not to enjoy the intrinsic rewards of the activity.

Social psychological research has indicated that extrinsic rewards can lead to overjustification and a subsequent reduction in intrinsic motivation.

Extrinsic incentives sometimes can weaken the motivation as well. In one classic study done by green & lepper, children who were lavishly rewarded for drawing with felt-tip pens later showed little interest in playing with the pens again.

Self-control

The self-control of motivation is increasingly understood as a subset of emotional intelligence; a person may be highly intelligent according to a more conservative definition (as measured by many intelligence tests), yet unmotivated to dedicate this intelligence to certain tasks. Yale School of Management professor Victor Vroom's "expectancy theory" provides an account of when people will decide whether to exert self control to pursue a particular goal. Drives and desires can be described as a deficiency or need that activates behaviour that is aimed at a goal or an incentive. These are thought to originate within the individual and may not require external stimuli to encourage the behaviour. Basic drives could be sparked by deficiencies such as hunger, which motivates a person to seek food; whereas more subtle drives might be the desire for praise and approval, which motivates a person to behave in a manner pleasing to others.

By contrast, the role of extrinsic rewards and stimuli can be seen in the example of training animals by giving them treats when they perform a trick correctly. The treat motivates the animals to perform the trick consistently, even later when the treat is removed from the process.

Motivational Theories

The Drive Reduction Theory

There are a number of drive theories. The Drive Reduction Theory grows out of the concept that we have certain biological needs, such as hunger. As time passes the strength of the drive increases as it is not satisfied. Then as we satisfy that drive by fulfilling its desire, such as eating, the drive's strength is reduced. It is based on the theories of Freud and the idea of feedback control systems, such as a thermostat.

There are several problems, however, that leave the validity of the Drive Reduction Theory open for debate. The first problem is that it does not explain how Secondary Reinforcers reduce drive. For example, money does not satisfy any biological or psychological need but reduces drive on a regular basis through a pay check second-order conditioning. Secondly, if the drive reduction theory held true we would not be able to explain how a hungry human being can prepare a meal without eating the food before they finished cooking it.

However, when comparing this to a real life situation such as preparing food, one does get hungrier as the food is being made (drive increases), and after the food has been consumed the drive decreases. The only reason the food does not get eaten before is the human element of restraint and has nothing to do with drive theory. Also, the food will either be nicer after it is cooked, or it won't be edible at all before it is cooked.

Cognitive dissonance theory

Suggested by Leon Festinger, this occurs when an individual experiences some degree of discomfort resulting from an incompatibility between two cognitions. For example, a consumer may seek to reassure himself regarding a purchase, feeling, in retrospect, that another decision may have been preferable. Another example of cognitive dissonance is when a belief and a behavior are in conflict. A person may wish to be healthy, believes smoking is bad for one's health, and yet continues to smoke.

Need Achievement Theory

David McClelland's achievement motivation theory envisions that a person has a need for three things, but differs in degrees to which the various needs influence their behavior: Need for achievement, Need for power, and Need for affiliation.

Interests Theory

Holland Codes are used in the assessment of interests as in Vocational Preference Inventory. One way to look at interests is that if a person has a strong interest in one of the six Holland areas, then obtaining outcomes in that area will be strongly reinforcing relative to obtaining outcomes in areas of weak interest.

Need Theories

Need Hierarchy Theory

Abraham Maslow's hierarchy of human needs theory is the one of the most widely discussed theories of motivation. The theory can be summarized as follows:

— Human beings have wants and desires which influence their behavior. Only unsatisfied needs influence behavior, satisfied needs do not.
— Since needs are many, they are arranged in order of importance, from the basic to the complex.
— The person advances to the next level of needs only after the lower level need is at least minimally satisfied.
— The further the progress up the hierarchy, the more individuality, humanness and psychological health a person will show.

The needs, listed from basic (lowest, earliest) to most complex (highest, latest) are as follows:

— Physiological
— Safety
— Belongingness
— Esteem
— Self actualization
— Herzberg's two-factor theory

Two-factor theory

Frederick Herzberg's two-factor theory, aka intrinsic/extrinsic motivation, concludes that certain factors in the workplace result in job satisfaction, but if absent, lead to dissatisfaction. The factors that motivate people can change over their lifetime, but "respect for me as a person" is one of the top motivating factors at any stage of life. He distinguished between:

— Motivators; (e.g. challenging work, recognition, responsibility) which give positive satisfaction, and

— Hygiene factors; (e.g. status, job security, salary and fringe benefits) that do not motivate if present, but, if absent, result in demotivation.

The name Hygiene factors is used because, like hygiene, the presence will not make you healthier, but absence can cause health deterioration.

Alderfer's ERG theory

Clayton Alderfer, expanding on Maslow's hierarchy of needs, created the ERG theory (existence, relatedness and growth). Physiological and safety, the lower order needs, are placed in the existence category, while love and self esteem needs are placed in the relatedness category. The growth category contains our self-actualization and self-esteem needs.

Self-determination theory

Self-determination theory, developed by Edward Deci and Richard Ryan, focuses on the importance of intrinsic motivation in driving human behavior. Like Maslow's hierarchical theory and others that built on it, SDT posits a natural tendency toward growth and development. Unlike these other theories, however, SDT does not include any sort of "autopilot" for achievement, but instead requires active encouragement from the environment. The primary factors that encourage motivation and development are autonomy, competence feedback, and relatedness.

Broad Theories

The latest approach in Achievement Motivation is an integrative perspective as lined out in the "Onion-Ring-Model of Achievement Motivation" by Heinz Schuler, George C. Thornton III, Andreas Frintrup and Rose Mueller-Hanson. It is based on the premise that performance motivation results from way broad components of personality are directed towards performance. As a result, it includes a range of dimensions that are relevant to success at work but which are not conventionally regarded as being part of performance motivation. Especially it integrates formerly separated approaches as Need for Achievement with e.g. social motives like Dominance. The Achievement Motivation Inventory AMI (Schuler, Thornton, Frintrup & Mueller-Hanson, 2003) is based on this theory and assesses three factors (17 separated scales) relevant to vocational and professional success.

Cognitive theories

Goal-setting theory

Goal-setting theory is based on the notion that individuals sometimes have a drive to reach a clearly defined end state. Often, this end state is a reward in itself. A goal's efficiency is affected by three features: proximity, difficulty and specificity. An ideal goal should present a situation where the time between the initiation of behaviour and the end state is close. This explains why some children are more motivated to learn how to ride a bike than mastering algebra. A goal should be moderate, not too hard or too easy to complete. In both cases, most people are not optimally motivated, as many want a challenge (which assumes some kind of insecurity of success). At the same time people want to feel that there is a substantial probability that they will succeed. Specificity concerns the description of the goal in their class. The goal should be objectively defined and intelligible for the individual. A classic example of a poorly specified goal is to get the highest possible grade. Most children have no idea how much effort they need to reach that goal.

Douglas Vermeeren, has done extensive research into why many people fail to get to their goals. The failure is directly attributed to motivating factors. Vermeeren states that unless an individual can clearly identify their motivating factor or their significant and meaningful reasons why they wish to attain the goal, they will never have the power to attain it.

Models of Behaviour Change

Social-cognitive models of behaviour change include the constructs of motivation and volition. Motivation is seen as a process that leads to the forming of behavioural intentions. Volition is seen as a process that leads from intention to actual behaviour. In other words, motivation and volition refer to goal setting and goal pursuit, respectively. Both processes require self-regulatory efforts. Several self-regulatory constructs are needed to operate in orchestration to attain goals. An example of such a motivational and volitional construct is perceived self-efficacy. Self-efficacy is supposed to facilitate the forming of behavioural intentions, the development of action plans, and the initiation of action. It can support the translation of intentions into action.

Unconscious motivation

Some psychologists believe that a significant portion of human behavior is energized and directed by unconscious motives. According to Maslow, "Psychoanalysis has often demonstrated that the relationship between a conscious desire and the ultimate unconscious aim that underlies it need not be at all direct." In other words, stated motives do not always match those inferred by skilled observers. For example, it is possible that a person can be accident-prone because he has an unconscious desire to hurt himself and not because he is careless or ignorant of the safety rules. Similarly, some overweight people are not hungry at all for food but for attention and love. Eating is merely a defensive reaction to lack of attention. Some workers damage more equipment than others do because they harbor unconscious feelings of aggression toward authority figures.

Psychotherapists point out that some behavior is so automatic that the reasons for it are not available in the individual's conscious mind. Compulsive cigarette smoking is an example. Sometimes maintaining self-esteem is so important and the motive for an activity is so threatening that it is simply not recognized and, in fact, may be disguised or repressed. Rationalization, or "explaining away", is one such disguise, or defense mechanism, as it is called. Another is projecting or attributing one's own faults to others. "I feel I am to blame", becomes "It is her fault; she is selfish". Repression of powerful but socially unacceptable motives may result in outward behavior that is the opposite of the repressed tendencies. An example of this would be the employee who hates his boss but overworks himself on the job to show that he holds him in high regard.

Unconscious motives add to the hazards of interpreting human behavior and, to the extent that they are present, complicate the life of the administrator. On the other hand, knowledge that unconscious motives exist can lead to a more careful assessment of behavioral problems. Although few contemporary psychologists deny the existence of unconscious factors, many do believe that these are activated only in times of anxiety and stress, and that in the ordinary course of events, human behavior — from the subject's point of view — is rationally purposeful.

Intrinsic Motivation and Basic Desires

Starting from a studies involving more than 6,000 people, Professor Steven Reiss has proposed a theory that find 16 basic desires that guide nearly all people behavior. The desires are:

- Acceptance, the need for approval
- Curiosity, the need to think
- Eating, the need for food

— Family, the need to raise children
— Honor, the need to be loyal to the traditional values of one's clan/ethnic group
— Idealism, the need for social justice
— Independence, the need for individuality
— Order, the need for organized, stable, predictable environments
— Physical Activity, the need for exercise
— Power, the need for influence of will
— Romance, the need for sex
— Saving, the need to collect
— Social Contact, the need for friends (peer relationships)
— Status, the need for social standing/importance
— Tranquility, the need to be safe
— Vengeance, the need to strike back

In this model, people differ in these basic desires. These basic desires represent intrinsic desires that directly motivate people behaviour, and not aimed at indirectly satisfying other desires. People may also be motivated by non-basic desired, but in this case this does not relate to deep motivation, or only as a means to achieve other basic desires.

Controlling Motivation

The control of motivation is only understood to a limited extent. There are many different approaches of motivation training, but many of these are considered pseudoscientific by critics. To understand how to control motivation it is first necessary to understand why many people lack motivation.

Modern imaging has provided solid empirical support for the psychological theory that emotional programming is largely defined in childhood. Harold Chugani, Medical Director of the PET Clinic at the Children's Hospital of Michigan and professor of pediatrics, neurology and radiology at Wayne State University School of Medicine, has found that children's brains are much more capable of consuming new information (linked to emotions) than those of adults. Brain activity in cortical regions is about twice as high in children as in adults from the third to the ninth year of life. After that period, it declines constantly to the low levels of adulthood. Brain volume, on the other hand, is already at about 95% of adult levels in the ninth year of life.

Besides the very direct approaches to motivation, beginning in early life, there are solutions which are more abstract but perhaps nevertheless more practical for self-motivation. Virtually every motivation guidebook includes at least one chapter about the proper organization of one's tasks and goals. It is usually suggested that it is critical to maintain a list of tasks, with a distinction between those which are completed and those which are not, thereby moving some of the required motivation for their completion from the tasks themselves into a "meta-task", namely the processing of the tasks in the task list, which can become a routine. The viewing of the list of completed tasks may also be considered motivating, as it can create a satisfying sense of accomplishment.

Most electronic to-do lists have this basic functionality, although the distinction between completed and non-completed tasks is not always clear (completed tasks are sometimes simply deleted, instead of kept in a separate list).

Other forms of information organization may also be motivational, such as the use of mind maps to organize one's ideas, and thereby "train" the neural network that is the human brain to focus on the given task. Simpler forms of idea notation such as simple bullet-point style lists may also be sufficient, or even more useful to less visually oriented persons.

Some authors, especially in the transhumanist movement, have suggested the use of "smart drugs", also known as nootropics, as "motivation-enhancers". The effects of many of these drugs on the brain are emphatically not well understood, and their legal status often makes open experimentation difficult.

Converging neurobiological evidence also supports the idea that addictive drugs such as cocaine, nicotine, alcohol, and heroin act on brain systems underlying motivation for natural rewards, such as the mesolimbic dopamine system. Normally, these brain systems serve to guide us toward fitness-enhancing rewards (food, water, sex, etc.), but they can be co-opted by repeated abuse of drugs, causing addicts to excessively pursue drug rewards. Therefore, drugs can hijack brain systems underlying other motivations, causing the almost singular pursuit of drugs characteristic of addiction.

Applications

Motivation is of particular interest to Educational psychologists because of the crucial role it plays in student learning. However, the specific kind of motivation that is studied in the specialized setting of education differs qualitatively from the more general forms of motivation studied by psychologists in other fields. Motivation in education can have several effects on how students learn and how they behave towards subject matter. It can:

Direct behavior toward particular goals

— Lead to increased effort and energy
— Increase initiation of, and persistence in, activities
— Enhance cognitive processing
— Determine what consequences are reinforcing
— Lead to improved performance.

Because students are not always internally motivated, they sometimes need situated motivation, which is found in environmental conditions that the teacher creates.

At lower levels of Maslow's hierarchy of needs, such as Physiological needs, money is a motivator, however it tends to have a motivating effect on staff that lasts only for a short period (in accordance with Herzberg's two-factor model of motivation). At higher levels of the hierarchy, praise, respect, recognition, empowerment and a sense of belonging are far more powerful motivators than money, as both Abraham Maslow's theory of motivation and Douglas McGregor's Theory X and theory Y (pertaining to the theory of leadership) demonstrate.

Maslow has money at the lowest level of the hierarchy and shows other needs are better motivators to staff. McGregor places money in his Theory X category and feels it is a poor motivator. Praise and recognition are placed in the Theory Y category and are considered stronger motivators than money.

— Motivated employees always look for better ways to do a job.
— Motivated employees are more quality oriented.
— Motivated workers are more productive.

The average workplace is about midway between the extremes of high threat and high opportunity. Motivation by threat is a dead-end strategy, and naturally staff are more attracted to the opportunity side of the motivation curve than the threat side. Motivation is a powerful tool in the work environment that can lead to employees working at their most efficient levels of production.

Nonetheless, Steinmertz also discusses three common character types of subordinates: ascendant, indifferent, and ambivalent whom all react and interact uniquely, and must be treated, managed, and motivated accordingly. An effective leader must understand how to manage all characters, and more importantly the manager must utilize avenues that allow room for employees to work, grow, and find answers independently.

The assumptions of Maslow and Herzberg were challenged by a classic study at Vauxhall Motors' UK manufacturing plant. This introduced the concept of orientation to work and distinguished three main orientations: instrumental (where work is a means to an end), bureaucratic (where work is a source of status, security and immediate reward) and solidaristic (which prioritises group loyalty). Other theories which expanded and extended those of Maslow and Herzberg included Kurt Lewin's Force Field Theory, Edwin Locke's Goal Theory and Victor Vroom's Expectancy theory. These tend to stress cultural differences and the fact that individuals tend to be motivated by different factors at different times.

According to the system of scientific management developed by Frederick Winslow Taylor, a worker's motivation is solely determined by pay, and therefore management need not consider psychological or social aspects of work. In essence, scientific management bases human motivation wholly on extrinsic rewards and discards the idea of intrinsic rewards. In contrast, David McClelland believed that workers could not be motivated by the mere need for money — in fact, extrinsic motivation (e.g., money) could extinguish intrinsic motivation such as achievement motivation, though money could be used as an indicator of success for various motives, e.g., keeping score. In keeping with this view, his consulting firm, McBer & Company, had as its first motto "To make everyone productive, happy, and free." For McClelland, satisfaction lay in aligning a person's life with their fundamental motivations.

Elton Mayo found out that the social contacts a worker has at the workplace are very important and that boredom and repetitiveness of tasks lead to reduced motivation. Mayo believed that workers could be motivated by acknowledging their social needs and making them feel important. As a result, employees were given freedom to make decisions on the job and greater attention was paid to informal work groups. Mayo named the model the Hawthorne effect. His model has been judged as placing undue reliance on social contacts at work situations for motivating employees.

REFERENCES

Buss, D. M. 1995. "Evolutionary psychology: A new paradigm for psychological science." *Psychological Inquiry*, 6, 1-30.

Darwin CR. 1871. *The Descent of Man and Selection in Relation to Sex*, Hurst and Co.

Durrant, R., & Ellis, B.J. 2003. "Evolutionary Psychology". In M. Gallagher & R.J. Nelson (eds.), *Comprehensive Handbook of Psychology*, Volume Three: Biological Psychology (pp. 1-33). New York: Wiley & Sons.

Pinker, Steven 2002. *The blank slate: the modern denial of human nature*. New York, N.Y: Viking.

Richards, Janet C. 2000. *Human nature after Darwin: a philosophical introduction*. New York: Routledge.

Tooby, J. & Cosmides, L. 2005. "Conceptual foundations of evolutionary psychology". In D. M. Buss (Ed.), *The Handbook of Evolutionary Psychology* (pp. 5-67). Hoboken, NJ: Wiley.

Bibliography

Alcock, J. 2001. *Animal Behavior: An Evolutionary Approach.* Sinauer Associates Inc., Massachusetts.

Andersson M. 1994. *Costs of Mate Choice; Direct Benefits of Mate Choice. In Sexual Selection.* Princeton, New Jersey: Princeton University Press.

Bandura A. 1965. "Influence of models' reinforcement contingencies on the acquisition of imitative response". *J. Personal. Soc. Psychol.* 1: 589-595.

Bateman AJ. 1948. "Intra-sexual selection in Drosophila". *Heredity* 2: 349-368.

Bateson P. 1983. *Sexual Selection by Female Choice by Peter O'Donald. in Mate Choice.* Cambridge: Cambridge University Press.

Baum, W. M. 2005. *Understanding behaviorism: Behavior, Culture and Evolution.* Blackwell.

Bennet MR, Harris JWK, et al. 2009. *Early Hominin Foot Morphology Based on 1.5-Million-Year-Old Footprints from Ileret,* Kenya. Science 323(5918): 1197-1201.

Bernstein DA et al. 2003. *Psychology,* 6th ed., Houghton Mifflin Co.: Boston.

Carew TJ. *Behavioral Neurobiology: The cellular organization of natural behavior,* Sinauer Assoc. Inc. Sunderland, Massachusetts.

Chomsky N. 1975. *Reflections of Language.* New York: Pantheon Books.

Clutton-Brock TH. 1990. *Reproductive Success: Studies of Individual Variation in Contrasting Breeding Systems.* University Of Chicago Press.

Cranston, A. 1994. "Self-Directed Animal Development Digestion." *The Journal of Animal Education* (Vol. 65, No. 6)..Ohio State Press.

Darwin CR. 1871. *The Descent of Man and Selection in Relation to Sex,* Hurst and Co.

Fantino, E.; Stolarz-fantino, S.; Navarro, A. 2003. "Logical fallacies: A behavioral approach to reasoning". *The Behavior Analyst Today* 4: 109–17.

Ferster, C. B., and Skinner, B. F. 1957. *Schedules of reinforcement.* New York: Appleton-Century-Crofts.

Greenberg N. 1977. *A Neurethological Study of Display Behaviors in the Lizard Anolis Carolinensis* (Reptilia, Lacertilia, Iguanadae). Amer. Zool. 17: 191-201.

Grier JW & T Burk, 1992. *Biology of Animal Behaviour.* Mosby Year Book, St. Louis.

Hailman JP. 1977. *Optical Signals: Animal Communication and Light.* Indiana Univ Press, Bloomington.

Hamilton, WD. 1964. "The genetical evolution of social behaviour I and II". *J. Theoretical Biology* 7: 1-16 and 17-52.

Konishi M. 1989. "Birdsong for Neurobiologists". *Neuron* 3: 541-549.

Kuhl P. 2000, *A new view of language acquisition.* PNAS 97(22): 11850–11857.

Lattal, K.A and Chase, P.N. 2003 .*Behavior Theory and Philosophy.* Plenum.

Menzel R, Mercer A. 1987. *Neurobiology and Behavior of Honeybees*. Springer-Verlag.

Merrin, L. 2007. *Animals: A Comprehensive Guide* (3rd ed.). Buffalo, NY: Bose-Jarman.

Menzer, J. 1991. *Animal Learning in the 21st Century*. Buffalo, NY: Bose-Jarman.

Mills, John A., 2000. *Control: A History of Behavioral Psychology*, Paperback Edition, New York University Press

Mulloney B, Selverston A. 1974. "Organization of the Stomatogastric Ganglion of the Spiny Lobster". *J Comp. Physiology* 91: 1-32.

Pfluger H-J, Menzel R. 1999. "Neuroethology, its roots and future". *J Comp Physiol* A 185: 389-392 pdf.

Pinker, Steven 2002. *The blank slate: the modern denial of human nature*. New York, N.Y: Viking.

Plotnik, Rod. 2005. *Introduction to Psychology*. Thomson-Wadsworth.

Rachlin, H. 1991. *Introduction to modern behaviorism*. 3rd edition. New York: Freeman.

Richards, Janet C. 2000. *Human nature after Darwin: a philosophical introduction*. New York: Routledge.

Scott, J.P. and Fuller, J.L. 1965. *Genetics and the Social Behavior of the Dog*. University of Chicago Press.

Skinner, B.F., *Beyond Freedom & Dignity*, Hackett Publishing Co, Inc 2002.

Snodgrass JG. 1975. "Psychophysics". In: *Experimental Sensory Psychology*. B Scharf. (Ed.) pp. 17-67.

Storey AE & DT Snow. 1990. "Postimplantation pregnancy disruptions in meadow voles: Relationship to variation in male sexual and aggressive behavior". *Physiology and Behaviour* 47: 19–25.

Tinbergen, Nikolaas 1963. *On the aims and methods of ethology*. Z. Tierpsychol. 20:410-433.

Tooby, J. & Cosmides, L. 2005. "Conceptual foundations of evolutionary psychology". In D. M. Buss (Ed.), *The Handbook of Evolutionary Psychology* (pp. 5-67). Hoboken, NJ: Wiley.

Verbeek, Peter. 2008 "Peace Ethology." *Behaviour* 145, 1497-1524.

Williams GC. 1966, *Adaptation and natural selection: a critique of some current evolutionary thought*. Princeton, N.J: Princeton University Press.

Wilson EO. 2000, *Sociobiology: The New Synthesis*, Twenty-fifth Anniversary Edition, Belknap Press.

Index